Care of Australian Reptiles in Captivity

John Weigel

"Every divergence deserves to be cherished, simply because it widens the bounds of life."
Karel Capek

Published by the

Reptile Keepers Association

To Robyn

Reptile Keepers Association
Box 227 Gosford New South Wales 2250

First published 1988

Weigel, John Randall, 1955-
Care of Australian reptiles in captivity.
Index 1. Reptiles. I. Title
ISBN 0 7316 3290 7

Printed by Clark Street Business Services, Crows Nest.

Reprinted 1989

Contents

Preface

Care of Australian Reptiles in Captivity is published by the Reptile Keepers Association (RKA), and its production has largely been a group effort. Editorial assistance was provided by numerous members of the RKA, officers of the New South Wales National Parks & Wildlife Service (NSW NPWS) and other amateur and professional reptile keepers. In particular, Glenn Shea and Gerry Swan of the RKA; Dr. Leighton Lewellyn, Principal Wildlife Management Officer, NSW NPWS; and Chris Banks, Curator of Reptiles at Melbourne Zoo have all contributed substantially of their time and expertise to make its completion possible.

Further valuable editorial comment was provided by Terry Boylan, Head of Reptile Section, Taronga Park Zoo; Peter Brinsley, NSW NPWS; John Cann, RKA; Neil Charles, RKA; John Montgomery, RKA; Dr. Keith Muelett, NSW NPWS; and Gary Stephenson, RKA.

Text, illustrations, colour photographs, design and wordprocessing are by the author, with the exception of the individual species care sheets, which were prepared by John Weigel, Glenn Shea, Gerry Swan, Grant Husband, Gary Stephenson and Ged Hughes - all of the RKA. The diseases section was co-written by the author and Glenn Shea (BVSc). Dr Harold Cogger kindly provided the excellent black and white identification photographs.

For encouragement, inspiration and endless support for amateur herpetology generally, the author and the RKA wish to thank Dr. Richard Shine. Special thanks are extended to Chris Chapman, who worked so hard to bring the hobby of reptile keeping to the standard it deserves. And finally, a thankyou to the wildlife officers of the various States and Territories who provided information on regional legal restrictions.

Laser printing and final wordprocessing by Jill Rassack of Silicon Quill, Hornsby.

Introduction

Reptile keeping in the home can be a fascinating experience for the whole family with a number of advantages over the keeping of more conventional pets. Relatively little space is required; accommodation can be aesthetically pleasing - and the reptiles themselves are clean, quiet and non-demanding, given a few simple requirements.

Care of Australian Reptiles in Captivity is intended to provide the novice with the basic information required for keeping reptiles in a healthy and well adjusted state. It provides step by step instructions on various aspects of the hobby, from building proper caging facilities, to modern reptile breeding techniques. It does not provide information on the keeping of crocodiles, marine reptiles, or highly specialized species as these are regarded as unsuitable for the beginner.

Recommended procedures for the safe handling and maintenance of venomous snakes are discussed, though the keeping of these species is not recommended for the novice or for persons under the age of 18. This subject is included in recognition of the fact that the majority of Australian snake species are venomous, and because many of these are widely kept by hobbyists at present. It is not the intention of the Author nor the Reptile Keepers Association to promote the handling or keeping of venomous species herein. However, if venomous snakes are to be kept, they should be maintained in a safe manner and in secure facilities. One of the goals of this book has been to provide adequate information on how these conditions can be met, with a section dedicated to the topic.

The popularity of reptile keeping is reflected in the burgeoning membership of reptile clubs and societies. It parallels an increasing regard that Australians have for their natural heritage and has resulted in a general loosening of legal restrictions to accommodate hobbyists. Of course, the legal protection of reptiles still extends to most states and territories, though in most of these the maintenance of at least a few reptiles by the hobbyist is recognized as an educational process, helping to break down levels of misunderstanding and fear about reptiles in the general community, and providing the impetus for many young enthusiasts to pursue scientific study.

However, with privileges are attached responsibility; the onus is on the reptile keeper to ensure that his or her charges are given the care and attention that they deserve. Hopefully, this book will provide assistance. It was initially conceived as a 'how to' book for New South Wales hobbyists, but its applicability to other States and Territories soon became apparent and its format was expanded to suit.

A selected list of useful references is provided should further information about reptile husbandry (or herpetology generally) be desired. Details of identification, distribution and biology of Australian reptile species can be found in the books listed. Additional queries can be directed to any of the herpetological associations listed below or the appropriate wildlife authority within your state or territory.

Table 1 A list of existing amateur herpetological associations in Australia. It is strongly recommended that reptile keepers belong to one of these groups.

Australian Herpetological Society PO Box R-79, Royal Exchange, Sydney, NSW, 2000	Reptile Keepers Association PO Box 98, Gosford, NSW, 2250
North Queensland Herpetological Society c/- 1 Faraday Street, Wulguru, Townsville, Qld, 4810	South Australian Herp. Group c/- South Australian Museum, N. Terrace, Adelaide, SA, 5000
Southeast Queensland Herpetologists c/- 9 Grey Street, Ipswich, Qld, 4305	Cape York Herpetological Society c/- Wild World, Box 88, Cairns, Qld, 4870
Western Herpetology Group 18 Creber Street, Whyalla Playford, SA, 5600	ACT Herpetological Group c/- ACT Parks and Conservation, PO Box 158 Canberra City, 2601

Legal Requirements

Before any attempt is made to collect or otherwise obtain reptiles, it must be understood that they are all protected by law. These laws generally carry with them substantial fines and penalties for non-compliance. The basic legal requirements vary considerably from state to state and are outlined below.

NEW SOUTH WALES

Note: At the time of reprinting the following legal requirements have not yet been formally adopted by the New South Wales National Parks and Wildlife Service. Before obtaining any reptiles, hobbyists should contact the Service for advice.

The relevant wildlife authority in New South Wales is: National Parks and Wildlife Service, Box 1967, Hurstville, 2220. Tel. (02) 585 6444.

All reptiles are protected in New South Wales under the National Parks and Wildlife Act 1974. There are substantial penalties for the illegal killing, capturing, holding and trading of reptiles. There are even higher penalties for offences involving reptiles classified as "endangered".

Exceptions to this blanket of protection allow the keeping of reptiles which have been "exempted" under Section 98(4) and Section 101(3), which may be collected from the wild (but not from areas managed by the National Parks and Wildlife Service). Additional reptiles may be held by persons licenced under the National Parks and Wildlife Act.

New South Wales Exempted Reptile Species

Specific caging requirements for each exempted species are outlined in the corresponding 'care sheets' that are contained within this booklet.

The makeup of the exempted reptile species list and the number of specimens of these that can be kept by each herpetoculturist is subject to periodic review by the Service. At the time of writing as many as 14 exempted reptiles (total number of specimens per keeper) may be held. In circumstances where reptiles in a person's collection breed it is permissible to hold the young for a three month period, even though the number of reptiles held for that period of time exceeds fourteen, without the necessity of applying for a Reptile Keepers Licence.

Table 2 'Exempt' reptile species for residents of NSW. These may be kept without licence in prescribed numbers and circumstances.

LIZARDS

GEKKONIDAE
robust velvet gecko - *Oedura robusta*
Tryon's velvet gecko - *O. tryoni*
broad-tailed gecko - *Phyllurus platurus*
thick-tailed gecko - *Underwoodisaurus milii*

PYGOPODIDAE
Burton's legless lizard - *Lialis burtonis*
common scaly-foot - *Pygopus lepidopodus*

VARANIDAE
sand goanna - *Varanus gouldii*

AGAMIDAE
jacky dragon - *Amphibolurus muricatus*
water dragon - *Physignathus lesueurii*
western bearded dragon - *Pogona vitticeps*

SCINCIDAE
land mullet - *Egernia major*
tree skink - *E. striolata*
White's skink - *E. whitii*
sandswimmer - *Eremiascincus richardsoni*
mountain water skink - *Eulamprus heatwolei*
eastern water skink - *E. quoyii*
pink-tongued skink - *Tiliqua gerrardii*
blotched bluetongue - *T. nigrolutea*
eastern bluetongue - *T. scincoides*
shingleback - *T. rugosa*

SNAKES

BOIDAE
carpet/diamond pythons - *Morelia spilota*

COLUBRIDAE
brown tree snake - *Boiga irregularis*
green tree snake - *Dendrelaphis punctulatus*

ELAPIDAE
common death adder - *Acanthophis antarcticus*
copperhead - *Austrelaps superbus*
yellow-faced whip snake - *Demansia psammophis*
Devis' banded snake - *Denisonia devisi*
swamp snake - *Hemiaspis signata*
tiger snake - *Notechis scutatus*
king brown snake - *Pseudechis australis*
spotted (blue-bellied) black snake - *P. guttatus*
red-bellied black snake - *P. porphyriacus*
western brown snake - *Pseudonaja nuchalis*
curl snake - *Suta suta*

TORTOISES

CHELIDAE
broad-shelled tortoise - *Chelodina expansa*
long-necked tortoise - *C. longicollis*
Macquarie tortoise - *Emydura macquarii*

Beyond the Exempted Species List: Getting Licenced

A person who wants to hold more than 14 exempted reptiles and/or any reptiles in New South Wales that are not on the "exempted" list must hold a Reptile Keepers Licence issued by the National Parks and Wildlife Service under Section 120 of the National Parks and Wildlife Act. A Reptile Keeper's Licence enables a person to hold more than 14 exempted reptiles that have been legally acquired. These reptiles may only be kept at premises specified on the licence.

The Reptile Keeper's Licence also allows the holder/keeper to acquire and dispose of non-exempt reptiles in a legal manner interstate. If you do decide to import specimens from another state or territory, you must apply for a NSW import permit, which requires a non-refundable fee of $10. Also, the supplier (other than Victorian residents) must tender an export permit from the appropriate authority of the state of origin.

All records of acquisition and disposal of fauna must be maintained in an approved manner. Licence holders will be required to notify the Service when animals are exchanged and when they breed, die, etc, and must pay an annual licencing fee.

Non-exempt reptiles may be acquired from any of the following sources: other licence holders who hold such reptiles legally; the wild (under the terms of a Scientific Licence); legal interstate sources; or as otherwise approved by the Service (e.g. injured animals or animals relocated from a residential area).

It must be clearly understood that the holder of a Reptile Keeper's Licence is not empowered to take non-exempt reptiles from the wild unless he or she has the specific approval of the National Parks and Wildlife Service to do so. Non-exempt species may only be collected under the terms of a Scientific Licence.

Scientific Licences

The National Parks and Wildlife Service is prepared to issue Scientific Licences for the collection of non-exempted reptiles from the wild. These Licences will not be issued lightly and applicants will need to propose bona fide research projects for this purpose. These projects need not necessarily be highly sophisticated ones, but they will need to be attempts to extend our knowledge and understanding of reptile fauna. A suitable project might be 'the study of the captive husbandry and propagation' of a particular species.

The Reptile Keepers Association and the Australian Herpetological Society will be required to oversee any activities carried on by their members under the terms of Scientific Licences. A condition of the Scientific Licence is that the licensee must submit an annual report to the Service outlining work undertaken under the terms of the licence.

Inspection of Licenced Reptile Collections

A condition attached to both the Reptile Keeper's Licence and the Scientific Licence to hold reptiles is that the premises where the animals are kept must be open for inspection by officers of the National Parks and Wildlife Service without prior notification. The Service has undertaken that these inspections will only be carried out between the hours of 9.00 am and 5.00 pm, and only by authorised officers of the Service. Such inspections will be conducted solely for the purpose of enforcing provisions of the National Parks and Wildlife Act.

Frogs As Snake Food

Since 1983 the following frogs have been fully protected against collection in NSW, and therefore *cannot* be collected as a food source for your snakes: green-thighed frog *Litoria brevipalmata*, *L. flavipunctata*, *L. glandulosa*, Jervis Bay tree frog *L. jervisiensis*, *L. spenceri*, *L. pearsoniana*, pouched frog *Assa darlingtoni*, Loveridge's frog *Kyarranus loveridgei*, sphagnum frog *K. sphagnicolus*, orange-crowned toadlet *Pseudophryne australis*, corroboree frog *P. corroboree*, and Fletcher's frog *Lechriodus fletcheri*. It is up to you to accurately identify any frogs that you may collect. H. G. Cogger's *Reptiles and Amphibians of Australia* is perhaps the best identification guide available [1986, Reed, Australia]. Also of value is *A Field Guide to Australian Frogs* by J. Barker and G. Grigg [1977, Rigby, Adelaide].

Any other frog species may be collected as food for frog-eating snakes, but the restrictions on the collection of exempted reptiles apply equally to the collection of frogs (i.e. not from nature reserves, state recreation areas, state forests or national parks; collecting to be done in non-destructive manner).

Exempted 'Feed Lizards'

Because many of the reptile species likely to be kept by hobbyists in NSW naturally prey upon small lizards, a list of suitable exempted 'feed lizards' has been compiled with the aim of providing an appropriate food source for certain predatory species held and to ensure that no rare or endangered lizard species is collected for food purposes. The following species may be collected for the purpose of providing food for lizard eating reptile species: *Gehyra variegata, Heteronotia binoei, Cryptoblepharus virgatus, Ctenotus regius, C. robustus, Hemiergis decresiensis, Lampropholis delicata, L. guichenoti, L. mustelina, Leiolopisma platynotum, Menetia greyii, Morethia boulengeri*

Catch And Release Licences

A limited number of licences are issued to private individuals allowing them to respond to requests from the general public (usually referred over the telephone from local police stations or Service offices) to have dangerously venomous snakes removed from private property. Details of the 'catch and release' program may be obtained at the Service's various offices.

Photographic Licences

Licences are issued on a case by case basis to private individuals allowing the capture and temporary holding of reptiles in NSW for photographic purposes.

Animal Rescue Program

From time to time, as the Service becomes aware of areas of bushland in NSW that are scheduled to be cleared for development, licences may be issued to reputable amateur herpetological groups to oversee the capture of certain, if not all of the reptiles occurring at the site. In some instances the Service may licence keepers to hold these specimens indefinitely, or direct that they be released in particular locations.

VICTORIA

The relevant wildlife authority in Victoria is: National Parks and Wildlife Division, 123 Brown Street, Heidelberg, 3084. Tel. (03) 450 8600.

The keeping of reptiles in Victoria is allowed. However, it is an offence to collect any reptiles within the state without a scientific permit, which is subject to a number of conditions. Therefore hobbyists generally must obtain reptiles from existing legally held collections within Victoria, or from interstate sources. In the case of the latter, no import permit is required, however, the Director must be notified within seven days.

All reptiles that are legally obtained need to be covered by an annually renewable licence or permit (depending upon the species) to keep reptiles, though the following species are exempt from this condition: long-necked tortoise *Chelodina longicollis,* Macquarie tortoise *Emydura macquarii,* marbled gecko *Phyllodactylus marmoratus*, thick-tailed gecko *Underwoodisaurus milii*, jacky dragon *Amphibolurus muricatus*, White's skink *Egernia whitii*, blotched bluetongue *Tiliqua nigrolutea*, eastern bluetongue *Tiliqua scincoides*, and shingleback *Tiliqua rugosa*. Remember that even the above listed species may not be collected from the wild.

The keeping of any native Victorian reptile species other than those listed above requires a licence with an annual fee attached. A record book is provided which must be kept as a proper record of transactions involving reptiles. It is legal to trade in reptiles where these are legally acquired.

SOUTH AUSTRALIA

The relevant wildlife authority in South Australia is the National Parks and Wildlife Service, GPO Box 1782, Adelaide, 5001. Tel. (08) 216 7777.

As of July 1988 there has been a new fauna permit system in South Australia. The new system is very complex, consisting of five classes of fauna, three classes of keep and sell permits and two classes of fauna dealer permits. Depending upon which class(es) of fauna are kept, licensees are liable for differing annual registration fees and due dates for the submission of returns (may be due quarterly, biannually, annually or not at all).

The new system places a limit on those species which may be kept on a class of permit but not on the total number of specimens which may be kept. Importantly, a second type of permit must be aquired allowing the actual collection of specimens from a natural environment. Not even those species of reptiles classified as 'exempted' may be taken from the wild without a permit. Clarification of the new system can be obtained by ringing or visiting the Fauna Permit Unit of the Service or by sending for a copy of the 16 page information booklet which contains lists of species catagorised in the various fauna classes.

QUEENSLAND

The relevant wildlife authority in Queensland is: National Parks and Wildlife Service, 138 Albert Street, Brisbane, 4000. Tel. (07) 227 5547.

All reptiles are protected in Queensland by relevant legislation which is enforced by the Queensland National Parks and Wildlife Service. Private reptile keepers are licenced by the Service to maintain their collections. This licence entitles the reptile keeper to maintain a number of reptiles at a premises and to 'deal' in up to 100 movements per year (one movement being one animal). The licence is renewable annually and a fee is payable at the time of application.

The regulations state that a permanent record must be kept of fauna held and of all transactions. This record is to be made available for inspection by a fauna officer upon request. Any reptiles moved must be accompanied by a 'Permit to Remove' which is available from the Service with a fee attached.

Reptiles are not often permitted to be taken from the wild for private keeping, but must be acquired from other registered keepers as captive bred offspring or as excess stock.

The reptiles held in a collection are regarded as always remaining the property of the crown.

WESTERN AUSTRALIA

The relevant wildlife authority in Western Australia is: Department of Conservation and Land Management (CALM), PO Box 104, Como, 6152. Tel. (09) 367 0333.

The attitude of the Western Australian government towards amateur herpetology differs significantly from all other Australian states and territories. It is regarded as a serious offence to keep any reptile for hobbyist purposes. Enquiring hobbyists should be aware that the detection and prosecution of unlicenced reptile keepers is rigorously persued by the Department CALM.

The only exception to this blanket prohibition exists in the licencing of one individual per wildlife region (eg one hobbyist only for entire Perth region) to hold certain venomous species to be used upon the direction of the Department of CALM in public education exhibits.

AUSTRALIAN CAPITAL TERRITORY

The relevant wildlife authority in the Australian Capital Territory is: ACT Parks and Conservation Service, PO Box 158, Canberra City, 2601. Tel. (062) 46 2849.

The keeping of reptiles in the Australian Capital Territory is allowed. However, it is an offence to collect any reptiles in the ACT or Jervis Bay Territory without a scientific permit, which is subject to a number of conditions. Therefore hobbyists generally must obtain reptiles from existing legally held collections within the ACT, or from interstate sources. In the case of the latter, an import permit is required.

All reptiles that are legally obtained need to be covered by an annually renewable permit to keep, though the following species are exempt from this condition: long-necked tortoise *Chelodina longicollis*, marbled gecko *Phyllodactylus marmoratus*, common grass skinks - *Lampropholis guichenoti*, *L. delicata*, and the eastern bluetongue *Tiliqua scincoides*. Remember that even the above listed species may not be collected from the wild.

Various classes of licences may be issued allowing the sale of reptiles where these are legally obtained.

TASMANIA

The relevant wildlife authority in Tasmania is: Department of Land, Parks and Wildlife, G.P.O. Box 44A, Hobart, 7005. Tel. (002) 30 2610.

Reptiles are not protected in Tasmania. Any non-endangered species may be collected and kept in any number from any area other than within the bounds of national parks. Reptiles may not be imported from other states without a permit from the relevant authority.

THE NORTHERN TERRITORY

The relevant wildlife authority in the Northern Territory is: Department of Forestry, Wildlife and Environment, National Parks Branch, Darwin, 5797.

At the time of writing, the wildlife regulations controlling the keeping of reptiles are undergoing change. It is anticipated that an exempt list of reptile species will be designated for collection and keeping by hobbyists.

Guidelines For Collecting Reptiles from the Wild State

In the event that reptiles are to be collected from the wild (eg exempted species or under permit by relevant authority), this should only be done in a sensible and responsible manner, and not at the expense of the environment. Under no circumstances should the collection of reptiles take place in national parks or other wildlife reserves. For many species (including most of the snakes and larger lizards), the most expedient means of collection is to drive along quiet country roads at the appropriate time of day or night. If your search is to be carried out on private property you must first obtain the permission of the property owner. There is no reasonable excuse for chopping down trees, using a pinch-bar to loosen and possibly damage rock exfoliations, or otherwise inflicting damage to the area you choose to carry out your search.

Keeping a written record of each specimen collected is of paramount importance. Such details as date of collection, species, time of day, location of capture, immediate habitat, weather conditions, body and tail lengths, and sex can be used as a basis for your captive records. You should also make note of your observations of the specimen while in captivity, such as rate of growth, quantity of food taken and frequency of skin-sloughing.

Housing for Reptiles

The captive environment for any reptile can be conveniently divided into various components such as cage type, heat, humidity, light, diet, space and the potential for interactions with other organisms and objects. However, one should keep in mind the fact that the captive reptile will not perceive its world in such neatly defined parameters, but rather as a whole, finding it either acceptable or unacceptable.

Outdoor Versus Indoor Caging Facilities

There are two very different types of housing for captive reptiles, the basic distinction being whether the reptiles are to be kept outdoors in fairly open circumstances, or in indoor cages, known more precisely as vivaria (singular vivarium). An advantage of the outdoor enclosure is that it can provide natural climatic and seasonal conditions for species that occur locally. On the other hand, indoor cages permit a greater degree of environmental control, and thus allow less hardy species or those that don't occur locally to be kept. Indoor housing can also result in the reptiles being more visible and therefore perhaps more likely to receive individual attention and care. Dangerously venomous snakes require special safety considerations. In New South Wales venomous snakes may only be kept within the security conditions set out on p 32.

Outdoor Enclosures

Outdoor reptile enclosures vary from aviary type pens to reptile 'pits'. The latter are not holes in the ground as the name might suggest, but smooth walled enclosures that may be entirely open above, or wire-topped to keep out small children and predators such as kookaburras, magpies and local dogs and cats. Pits should only be contemplated for maintaining tortoises and active, sun-loving lizards that naturally occur in climates similar to your own. Aviary style pens can be purchased or built to house certain of the larger lizards and non-venomous snakes. Transparent or opaque fibreglass roofs over a slightly raised floor provide a means of maintaining warmth and dryness, but care must be taken that a 'green house' effect doesn't arise causing excessive temperatures below. The important concepts of cage furnishings and provisions (ie: provision of water, hide boxes, clean substrate materials, etc) discussed in the indoor caging section of this booklet may also be applied to any outdoor pens used to house reptiles.

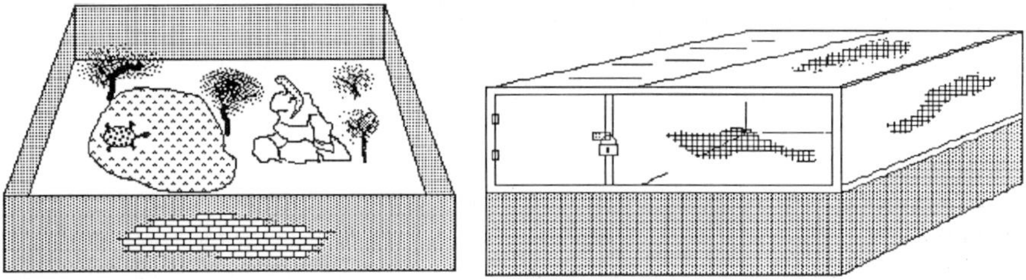

Figure 1 Two types of outdoor caging facilities for reptiles - a reptile pit *left* and an adapted aviary *right.*

Outdoor facilities that are large enough can house more than one species, provided they are compatible. The type and number of reptiles held in any outdoor facility is related to the size of the facility and the spatial requirements of the species housed. The enclosure's interior design should be consistent with the environmental needs of all inhabitants. All outdoor enclosures should be built in such a position and manner as to receive substantial periods of sunshine at all times of the year, yet constructed or decorated in such a way as to also provide adequate shade areas throughout the day. During the winter months, reptiles kept outdoors in temperate areas will probably enter a state of relative inactivity if not hibernation and must have access to suitable retreats. In the case of tortoises and some lizards, either large flat river stones, broad strips of bark, or a pile of dry vegetation should be available to burrow beneath, while other lizards may prefer dry, clean crevices between rocks or suitable hollow logs in which to spend the cooler months. Occasionally, tortoises will choose to overwinter within a pond, remaining submerged and taking in adequate oxygen through the skin.

Some keepers opt to move their reptiles into indoor caging facilities during the winter, either maintaining warm conditions, in which feeding is continued throughout, or cool, quiet, private conditions intended to simulate a natural inactive period. The importance of a cooling down period for reproduction of some species in captivity is discussed in the breeding section of this book.

Unfortunately, before building an outdoor caging facility for your reptiles, you will need to consider the possibility of theft or unwelcome visits from neighborhood children, cats and dogs or attacks by predatory birds that can kill or torment reptiles kept in pits. Pens can be made predator-proof and fitted with locks, but can never be made entirely theft-proof.

Reptile Pit Construction

As mentioned earlier and contrary to the implication of the name, reptile 'pits' should not be dug beneath the ground. Unless well drained, any areas of a pit below ground level are subject to flooding or muddy conditions following rains.

The retaining wall of your reptile pit can be constructed from formed concrete, rendered bricks, asbestos-type construction sheets or galvanized iron. A simple, relatively cheap alternative is to bolt or rivet one metre wide lengths of sheet metal or equivalent material together to form a circular enclosure. The minimum height of the pit wall will depend upon the size of the tortoises, the leaping ability of the lizards or the length of snakes to be maintained within. Careful attention should be paid to the enclosure to ensure a smooth, non-climbable barrier, particularly along any corners or joins. In the absence of a floor, the walls of your pit should continue deep enough into the ground to prevent the inhabitants from burrowing out, or wild rodents from digging in.

Trees and shrubs can add an aesthetic touch to any reptile pit, but if placed too close to the wall will allow escape. The same rule applies to any rock outcrops, hollowed logs, etc. that you choose to include. With a bit of imagination and some creative landscaping, you can produce an aesthetically pleasing facility that recreates to a large extent the natural environment of the species being maintained.

Pools for the Reptile Pit

Unless extenuating circumstances exist, fresh water should be available at all times in the reptile pit. Water bowls are not always practical if the facility is of substantial size and the inhabitants are unlikely to find their water readily. Therefore, you might consider including a sunken pool of some description. Pool construction in your pit will depend upon the sorts of reptiles being kept. Semi-aquatic lizards should be able to completely immerse themselves, and tortoises should have a pond with a minimum depth of 50 cm. Any pool or pond should have gently sloping sides on at least one end to allow easy access and exit by the reptiles.

A suitable pool for lizards can be established simply by sinking a plastic or metal wash-tub, old bathtub, or perhaps an infant's swimming pool into the ground with its upper edge flush or just above the ground level. Old concrete wash tubs are unsuitable as they slowly release accumulated toxic detergents into the water. You may need to leave a large stone or a water-soaked log in the pool to allow the inhabitants of the pit to climb out. Larger ponds intended for tortoises can be constructed by simply filling an appropriately sized hole in the ground with water (so long as local soil types are suited), lining the hole with a large sheet of heavy plastic or lining the pool with impervious clay.

Constructing the Tortoise Pond

The best pools for tortoises are constructed from concrete. The pond should be situated so that during both winter and summer the sun will reach it most of the day. It is important to consider the number of tortoises that are intended to be kept before building the pond, as overcrowding must be avoided. A volume of water about 2m x 2m x 1/2m (average depth) is adequate for three tortoises with shell lengths of approximately 30 cm. When digging the hole, remember to compensate in all dimensions particularly depth for the anticipated thickness of cement to be added. You will also need to make at least part of the pond with a gently sloping wall to allow the tortoises easy access and departure. After having excavated the initial hole the soil must be well tramped to avoid future settling. Where possible, a drain pipe should be inserted at the lowest point of the pond to feed to a lower point somewhere outside of the reptile pit, making future water changes and cleaning much easier. Steel reinforcement should be laid out along the walls and bottom of the pond before the cement is applied at an even thickness (about 75mm is best).

Before allowing the cement to dry, it is important to smooth the surface to avoid future abrasion to the tortoises' shells. After two days drying you will be able to scrape away any rough areas of the finished surface with a trowel. The concrete should then be sealed with Silasec, Cemstick, fibreglass resin or other suitable sealant according to the directions supplied with these products. Alternatively, certain products (e.g. Boncrete) can be mixed into the cement with the same result: making the pond smoother and water-tight. When the concrete is completely dry, water must be added to neutralize the extreme alkalinity of the cement, and allowed to sit for at least a week. The pond can then be drained and thoroughly scrubbed before rinsing once or twice again.

When finally filled, the pH level of the water must be tested and maintained at a neutral level pH 7. Deviation from a neutral pH condition predisposes tortoises to fungal diseases and other health risks. A pH testing kit, together with the chemicals required to adjust the pH level, may be purchased at most swimming pool supply shops and larger pet shops.

Vegetation around the pond creates a natural effect, provides shade and cover, and attracts insects (for food). The water should be occasionally changed or kept clean either by filtration or by a constant trickle of fresh water. Under-gravel filters are cheap and easy to adapt or make. Water plants provide a food supplement and cover for tortoises. Additionally, small tortoises will use floating leaves as basking sites. Mud or sand can be added to the pool's bottom to create a natural resting place for submerged tortoises.

Heating and Lighting Outdoor Enclosures

Theoretically, you shouldn't have to supply any artificial heating or lighting for reptiles kept in properly constructed outdoor caging facilities, so long as the species being kept are suited to the local climate. Sufficient sunlight must be available, and adequate attention must be paid to design as mentioned previously. Nevertheless, it can be beneficial to provide supplemental warmth to the most frequently used hiding spaces. It is possible to build an artificial 'basking site' within a large hide-box facility. Some of the heating devices which can be adapted for outdoor use are described in the section on heating the indoor terrarium.

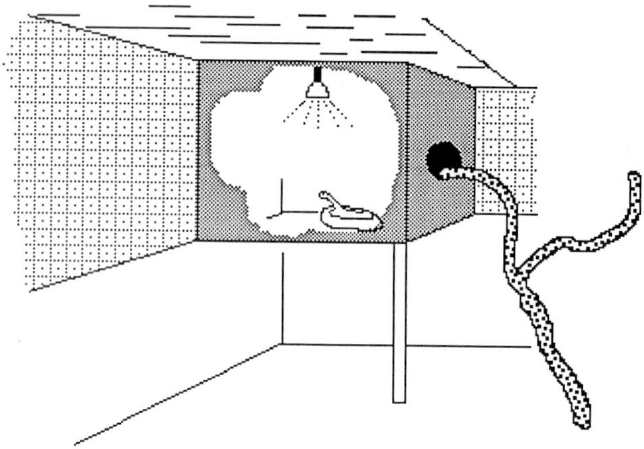

Figure 2 Cutaway view of a heated chamber in an outdoor enclosure. Access to the entryway is provided by a securely positioned tree branch.

Indoor Caging Facilities

For the majority of reptile keepers, the indoor vivarium offers a more suitable alternative than the outdoor facility. Vivaria include terraria (dry cages for predominantly terrestrial species), aqua-terraria (partly aquatic, partly dry cages) and water-filled aquaria (for indoor maintenance of tortoises).

Aquaria and Aqua-Terraria for Maintaining Tortoises Indoors

Aquaria and 'aqua-terraria' can be used to house young tortoises for extended periods indoors, though in the case of aquaria, a means for periodically drying off completely must be provided - i.e. a log or stone must protrude from the water's surface. An aqua-terrarium contains varying amounts of both land and water, allowing the tortoise the choice of leaving the water to dry or bask. Unless positioned so as to receive sufficient direct sunlight through a window, indoor tortoise keeping facilities will require an artificial source of ultraviolet light. Ideally a filtration system should be installed, or else the water will need to be exchanged fairly often. Ideal water temperature would be around the 25C mark. Placing water plants in the tank can be a good idea, though some varieties are unfortunately short lived with tortoises, as what doesn't get eaten will be scratched out by the roots. Careful attention to maintaining water at a neutral pH 7 will reduce the possibility of diseases.

There are various ways of constructing an aqua-terrarium but the simplest method is to take a water tight aquarium and use a silicone rubber sealant to glue a smooth edged piece of glass width-wise across the middle of the tank bottom with a height of say, 20cm (see diagram). This will create a division, so that water can be added to the one half, with the rest of the cage committed to the creation of a 'dry-land' environment.

The Terrarium - Indoor Caging for Snakes and Lizards

Terraria can be constructed from any of a wide variety of building materials, so long as adequate attention is paid to making the end-product completely escape proof. Of course, what is escape proof for one species may be easily departed by another. Bluetongues and other large lizards can be kept in open-topped glass tanks of sufficient height. Snakes on the other hand are able to squeeze through the smallest of gaps or holes, and can even force their way through flimsy air vents, doors or lids. Small specimens can wet themselves in the water bowls provided and take advantage of the surface tension afforded by their dampness to scale tall glass cages. As a rule glass viewing fronts, rather than flyscreen ones are preferred for the terrarium since reptiles can scrape their snouts on flyscreen trying to get past it.

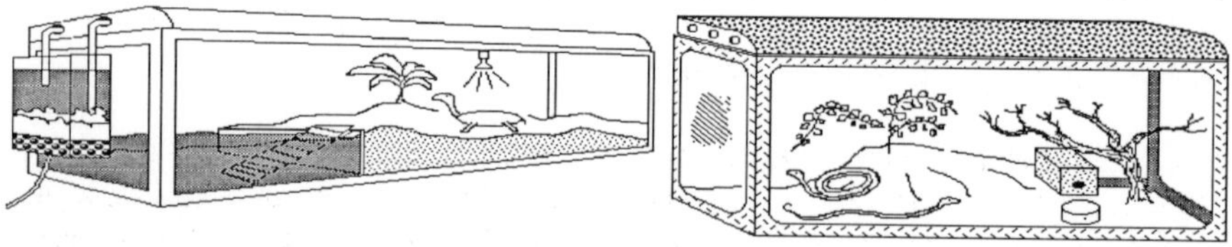

Figure 3 An aqua-terrarium *left*, and a terrarium *right*.

The simplest container for the beginner is a glass or plastic tank of the type sold as aquaria. Depending upon the species being kept, a well ventilated lid may need to be fitted which may also support the heating and lighting equipment. Old wooden cabinets can often be purchased cheaply at second hand furniture shops, and can be fitted with air vents and possibly twin sliding glass doors to build a very satisfactory terrarium.

If more than one cage is required, a popular and cost effective method is to incorporate a series of cages as a unit, perhaps one or two cages across and four to six cages high, extending from floor to ceiling. Such a 'bank' of cages can easily be built using cheap, commercially available book shelving units as the basic foundation. A sheet of plywood can be attached to the front of the unit, from which hinging doors with glass viewing centres can be cut with a jig-saw. Alternatively, sliding glass doors with plastic runners can be installed into the fronts of the separate compartments.

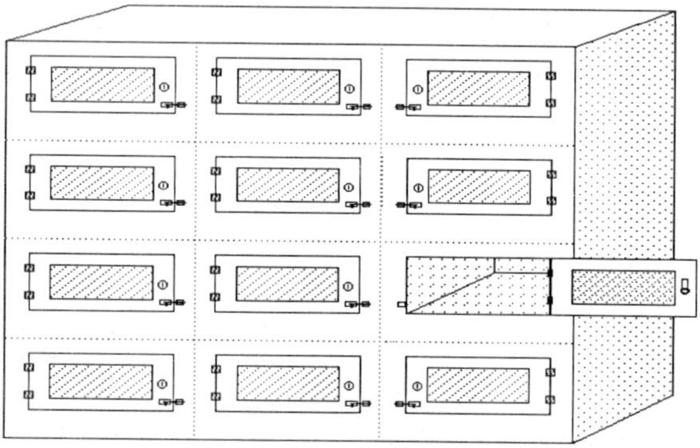

Figure 4 Banks of cages may be constructed to house numerous reptiles in a relatively small area of the home. These may require a peg-board back or equivalent for ventilation and to avoid excessive heat build-up.

This system allows the fitting of a number of cages into a small area with the added advantage of more efficient cage heating. For example, a light globe in one cage will also heat at least part of the floor of the cage above. The value of 'floor heating' is discussed in the section on cage heating.

The interior surface of any cage (particularly the floor) should be smooth and water resistant so that it can be cleaned. For wood or particle-board cage interiors, a couple of coats of paint or a sealant such as polyurethane should be applied. If spirit-based products are used within cages, you will have to wait for the fumes to dissipate before introducing the reptiles.

You should occasionally empty your cages entirely and scrub them down with a disinfectant such as an ammonia or chlorine solution. Always follow the safety precautions printed on the bottles and rinse well. Don't use 'pine-o-clean', eucalyptus oils or any other products that leave strong and lingering odours unless they are used in glass tanks that can be rinsed well and put out in the sun for a day.

Substrate Choices for the Terrarium

Potentially suitable floor coverings include natural materials such as a shallow base of coarse river or beach sand with a leaf-litter covering. The exfoliated bark of Eucalyptus and Angophora trees provide an excellent choice for small secretive snakes, creating plenty of cover to hide beneath. Some keepers prefer the sanitation and simplicity offered by the use of newspaper or 'butchers' paper as a primary floor covering, with or without a layer of leaf-litter or bark over it. If the terrarium is to be used for small lizards, it is recommended that you stay away from aquarium gravel or similar gravel floor covering since these lizards have been known to ingest large amounts of the tiny stones, leading to intestinal blockage and death. Regardless of the sort of substrate you choose, it must be kept dry, free of faeces and uneaten food, and at least occasionally replaced. Sand and stones may be washed and heat-dryed.

Water Requirements in the Terrarium

Under most circumstances fresh water should be provided on a regular basis from either a non-spillable container or a built-in pool. In the case of arid-dwelling species or a gravid (pregnant) female that is about to lay eggs you may choose to only provide water on a daily or weekly basis, removing the bowl after you have observed all captives drinking their fill, or to restrict access to the built in pool. Captive reptiles have been known to lay their eggs in any available standing water, which will kill the embryos within.

Relative Humidity

The relative humidity within any caging facility is influenced by the presence of drinking water and live plants within, the internal air temperature, the rate of air exchange from outside the cage, and the relative humidity and temperature outside the cage. Too high a humidity may predispose your reptiles to skin afflictions or respiratory problems, while too little may result in dryness of skin or dehydration which can lead to further complications. The best indicator of how much humidity your animal needs will be found in a careful study of its natural history. Desert species that are active by day usually do best in dryer conditions than those which would be suited to coastal and montane species.

Plants and Other Cage Furnishings

With the aid of natural materials and plants it is possible to decorate the inside of a terrarium so that it will be an attraction in any room. If the reptile normally resides in rock crevices, you can create a miniature rock outcrop within the cage, though it must be very stable to avoid becoming a death trap. If too elaborate however, extraction of the specimens from within will be difficult. Tree climbing species can be accommodated by the inclusion of a suitable branch. A desert scene, or perhaps a miniature rain-forest setting can be constructed by including appropriate natural materials and possibly plants. As previously mentioned, if you intend using live plants in the terrarium, be careful that the relative humidity does not become excessively high, and that moist potting soil doesn't become an incubator for disease agents. With sufficient thought, foresight, and experimentation, you can come up with a natural arrangement that not only looks good, but is well received by the terrarium occupants.

The Importance of Hiding Places in the Terrarium

Most reptiles will adapt more quickly to captivity if suitable cover, or 'hiding spaces' are provided. In the wild, most snakes and lizards spend an inordinate amount of time in confined and even tight fitting recesses, such as in burrows, beneath flat stones, or in hollow bits of timber. In captivity, the importance of providing such privacy cannot be overemphasized. 'Hide boxes' have always been a popular source of such refuge for captive snakes, and can be made from discarded cardboard boxes of appropriate size, commercially constructed bird breeding boxes, etc. Bird boxes can be water-proofed within and provide ready access to the reptile since their tops are either hinged or removable. They are available in a variety of sizes, and are sold at pet shops everywhere. If you use them, it is a good idea to make sure that the access hole is sufficiently large, so that your animals cannot become trapped if one decides to leave the box while another enters. If there is any doubt, enlarge the hole.

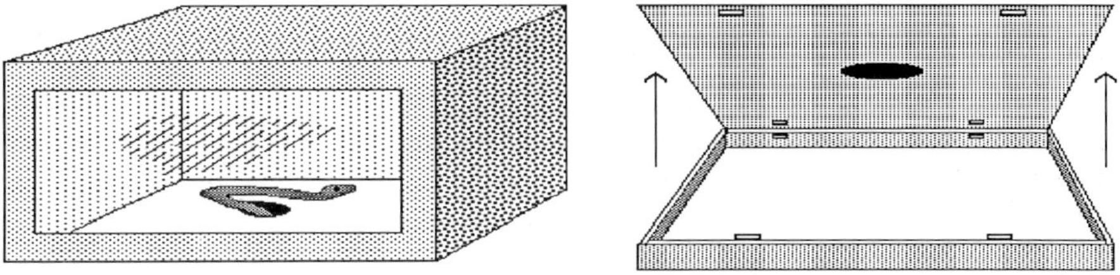

Figure 5 A 'false floor' can provide a dark retreat for reptiles while effectively doubling the cage floor space. The removable floor should be fastened when in place with magnetic cabinet latches or equivalent.

Another way of creating hide space is the use of 'false-bottomed' cage floors - which are most suited to species that naturally spend a great deal of time in underground retreats. The false-floor of course will need to have a small entrance hole somewhere in it, and the floor should be removable for ready access to the reptile and for ease of cleaning beneath. The space below the false-floor should be dark. By providing a false-floor, you not only provide the snake with a spacious, darkened hide space, but also effectively double the floor space of the cage. Special consideration to heating is necessary though, since any heat source directed from above is unlikely to contribute substantially below the false-floor. For this reason a floor heater should be fitted beneath the cage or at least beneath the darkened hide space (see section on cage heating).

One final word on hide spaces: whatever type is used, it is important that it can be disassembled rapidly for access to any reptiles within. Therefore, long and narrow hollow logs, or similarly awkward hiding cavities are not considered to be suitable.

Heat and Light for the Vivarium

Depending upon the species being kept, a reptile's lighting and heating needs might be at least partially met by close proximity to a sunlit window (though direct sunlight onto glass-fronted vivaria should be avoided since extreme temperature rises could result). As a rule however, indoor reptile cages must be provided with artificial light and heat. This is normally accomplished from within the cages themselves. Heating the room in which the cages are kept as a sole means of heating the reptiles is really only practical where a large number of cages are kept in a fairly small room.

Cautionary Notes Regarding Electricity

Consult an electrician before attempting to install any electrical lighting or heating devices. For all but the simplest procedures, you are required by law to have an electrician carry out this work. It is a good idea to ensure that your reptiles cannot burn or electrocute themselves by directly contacting any such devices. You may need to surround heating or lighting fixtures with wire mesh or similar 'reptile-proof' housing material.

Artificial Light Sources

Artificial lighting can be produced using either fluorescent tubes or incandescent light globes. An advantage of using fluorescent lights is the availability of special tubes that are high in the transmission of ultraviolet (UV) light, which is important in the absorption and synthesis of certain vitamins and minerals, particularly by diurnal lizards. Probably the best fluorescent tubes in terms of UV light production are the 'black light' tubes - not to be confused with the 'psychedelic' black lights, nor bactericidal lights. Some types can cause severe burns to reptiles. The most widely used brand in vivaria is the 'True-lite', also called 'Power-twist'. It has a reasonably long life and produces a balanced 'daylight' colour. The 'National Black Light FL 20SS-BL/8' and the 'NEC T-10 Blacklight' offer alternatives which produce greater amounts of desired UV light than the True-light, while costing somewhat less. All UV light producing tubes should be replaced every six to twelve months. Special 'gro-light' incandescent globes are also available that are intended to mimic sunlight in order to grow plants indoors, though none of these globes has been shown to match the ultraviolet light transmission of some of the fluorescent tubes currently on the market. When using artificial ultraviolet light sources, it must be recognized that any glass or plastic materials placed between the light source and the intended recipient of the ultraviolet light will filter some if not most of the ultraviolet rays out of the light. This applies even to the opaque plastic filters that are usually supplied with fluorescent light fixtures.

In order to adequately light the interior of say, a one metre long terrarium, an internally mounted 18" or 24" fluorescent light fixture is ideal. If an incandescent globe is to be used to light a similar sized set up, anything from 25 watts to 100 watts might be suitable, depending upon thermal requirements, which are discussed below.

Why Heating is Necessary

Although commonly thought of as 'cold-blooded', reptiles prefer to maintain their body temperatures at a level similar to many mammals. The cannot contribute appreciably to their own body temperatures the way mammals do, but rather absorb heat from environmental sources. Reptile activities such as feeding and digestion generally take place within a narrow range of preferred body temperatures.

Various behavioral strategies are employed by the different species to fulfill their thermal requirements in the wild state. Many diurnal species spend a considerable amount of time sun basking, while some nocturnal or small species shuttle between sun-warmed rocks or other materials during the early hours of the night in order to maintain an optimal body temperature. Other species are able to function with fairly low body temperatures. You will need to recognize and accommodate the individual thermal requirements of whatever species you are keeping. If kept at the wrong temperature, your reptiles might look healthy and may even feed voraciously, but will not be able to properly digest their food, and will become susceptible to potentially fatal diseases. Many species can withstand extended periods below the preferred 'optimal' temperature range under natural conditions, but under captive conditions where stress may be a factor, extended maintenance of reptiles at sub-optimal temperatures is to be avoided since it can result in a breakdown of their defences against disease. The notable exception to this rule is in the provision of a winter 'cooling down' period required to bring some species into breeding condition. However, this must be done in a carefully prescribed manner (see breeding section). How you can best provide heating for your reptiles will depend upon the species being kept - you will have to refer to the care sheets at the end of this booklet to determine the temperature requirements for each.

It is a failure to provide adequate warmth for their charges that has proven to be the undoing of many novice reptile keepers in the past. An all too common failing has been an inability to recognize that the floor temperature of a cage might be significantly lower than the walls or upper portion of the cage (where a thermometer reading might be taken), especially if the cage is kept on a cold floor and is heated solely from above. Therefore, temperature readings should be taken with a thermometer from various positions within the cage, including in any hide boxes or other 'hiding places' where the reptile spends substantial amounts of time, before judging whether or not suitable heating is being provided. In the end, it is the reptile's body temperature, as influenced by the immediate environment, that must be kept within critical limits, *not* the general cage temperature. 'Optimal' body temperatures vary between reptile species, but can be accommodated by creating a 'thermal gradient' across the floor of the terrarium.

The Importance of Creating a Thermal Gradient

It is known that different reptile species vary considerably in terms of their thermal requirements and corresponding preferred body temperatures. Additionally, this 'preferred body temperature' can be variable for any individual reptile depending upon such factors as whether or not it has recently fed; if it is about to begin skin sloughing; time of the day; health; etc. It is therefore essential that you provide a range of temperatures within the terrarium at any given time so that the reptiles may move within the vivarium to control their body temperatures and therefore carry out their various life functions in an optimal manner. As an example, immediately after a reptile feeds it will typically seek out a relatively warm spot so that its digestive enzymes are able to function properly. On the other hand, a reptile may choose to seek a cooler spot in the terrarium. In providing a 'thermal gradient', you are also ensuring that your reptiles will be able to withstand the effects of extreme periods of hot or cold weather until temperature conditions return to normal, since there is likely to be a position within the vivarium that remains 'cool enough' or 'warm enough'. A sensible gradient for most reptile species would be a gradual change from a relatively cool position of 24C, to a high of 30C, though a warmer reading would of course occur directly beneath any basking lamp.

Using Light Globes to Heat Cages

Diurnal (active by day) species that normally bask in the sun generally have fairly high heat requirements that are well suited by an internally mounted reflective 'spot-globe'. The globe can be directed onto some portion of the cage floor, or perhaps upon a branch, stone or other potential basking place. These are available in various shapes, sizes and wattages in specialist lighting shops. You may have to experiment with a suitable selection of globes before finding one with the right amount of heat and light production. Beware of reflector globes that concentrate their light onto an especially small area as this can result in skin burns to basking reptiles. Experiment by putting your hand in the most directly lit position - the exposed skin of your hand should become no warmer than if it were exposed to the sun itself on a summer day. If you decide upon a reflector globe or spot globe, the light should directly cover an area of approximately half the cage floor, with no exceptional 'hot spots' in the centre of the lit area. Large cages may require more than one globe.

Using Coloured Globes to Provide Night-time Heat

Reptiles should be provided with a natural regimen of alternating 'day' and 'night' cycles of corresponding lightness and darkness. This is easily accomplished outdoors, but can be more difficult for indoor vivaria that are heated by conventional light globes. Of course, one could argue that if the species being kept occurs locally, auxiliary heating/lighting shouldn't be necessary at night and the light globe can simply be turned off. In truth however, at night, as by day, there is a cline of temperatures in any given natural environment, and reptiles are able to seek out and occupy comparatively warm points of refuge (eg. under a sun-warmed rock that retains heat throughout much of the night). In confinement, where no such warm spots exist, a reptile may lack sufficient warmth.

One solution to this problem of heating the cage while preserving 'darkness' is to provide a second light globe fitting - powered independently and fitted with either a red or blue coloured globe. This second globe should be turned on at night when the conventional light globe is turned off, and may be of a somewhat lesser wattage to allow some degree of cooling at night. Blue or red light globes provide reptiles with a false impression of twilight or near-dark conditions while generating warmth. The added advantage in using a coloured light globe at night is that it will allow you to watch the behaviour of geckos and other nocturnal species during their peak activity periods. Avoid coloured globes that have a transparent glass 'neck', because the white light that will be emitted from this uncoloured section will defeat the purpose.

Heat resistant automobile engine paint is available in red, which can be sprayed onto standard globes and reflector lamps to create suitably coloured light at considerable savings.

Heating From the Cage Floor

An alternative or supplementary means of heating the vivarium, is to warm the cage floor from below. This can be a more cost-effective way of providing the necessary warmth for cage inhabitants, though the initial cost of purchasing or constructing a suitable heating device must be considered. Check with an electrical supplies shop for the details as to what can safely be installed. 'Pet warmers' designed to fit in dog and cat boxes are available through pet shops and are a useful option. Electric blankets have been successfully used when placed beneath cages, but under no circumstances should they be folded.

If cages are built in a modular manner (one above another), the primary heat source within each cage will create a warm spot on the floor of the cage above it. Of course this doesn't solve the problem of heating the floor of the bottom cage! To accomplish this, as yet another alternative, a subfloor compartment can be fitted with a light fitting, preferably to be positioned towards one end of the floor so as to produce a temperature gradient across it.

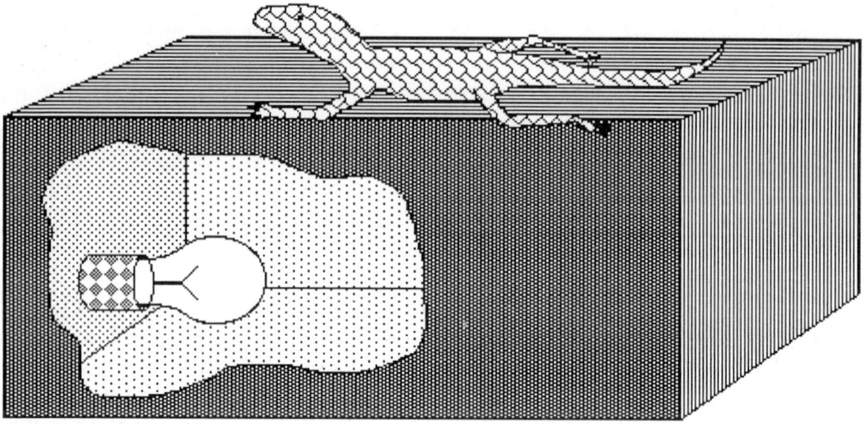

Figure 6 Cutaway view of a wooden 'hot box' with a 15 watt oven globe mounted towards one end to create a thermal gradient across top. Wood-screw fastening of one side allows changing of globes. One of the sides (not top) should be of pegboard to avoid overheating within.

Another popular means of providing auxiliary heat from below is to build a 'hot box' out of timber that encloses a fitting for a low output globe (15-25 watts depending upon size of box). This will create the equivalent of a sun-warmed stone within the terrarium, and a place for the reptiles to warm themselves whenever they choose. Make sure that the surface temperature of the box is not too high for the reptile species being maintained, or it will not be used. Peg-board or an equivalent material should be used on at least one side of the box (not top) to prevent overheating within, and some form of access (such as a side which can be removed by a few wood-screws) must be provided for changing burnt out globes.

Thermostats and Rheostats

Thermostatic control of the terrarium's heat source is advised in order to maintain steady temperatures and to avoid accidental overheating. The thermostatic control can be set to control the desired temperature of its position (or its remote sensing bulb if it has one) in the terrarium. The position of its placement should be where an average or below average temperature reading for the cage is likely to occur. The thermostatic bulb or sensing device should be shielded against direct exposure to any heating globes. Such direct exposure will result in the actual cage temperature being kept too cool since the thermostat will continuously cut the heat source off prematurely.

Notwithstanding the value of thermostats, a fair degree of temperature control can also be achieved by using a rheostat, or 'dimmer switch', that allows either floor or globe heating output to be adjusted precisely. However such a system is of course insensitive to weather and temperature changes and therefore the risk of overheating still exists if it is not used in conjunction with a thermostat. If you intend to install either a thermostat or rheostat, consult an electrician for advice and assistance.

Heating and Lighting Indoor Tortoise Facilities

The addition of an electric tube-type aquarium heater is ideal for heating the aquarium or aqua-terrarium. Most have a built-in thermostat, and have variable settings to allow the maintenance of an even, desired water temperature. A water temperature of about 25°C will normally suffice. A water thermometer should also be added as a precautionary measure. A fluorescent ultraviolet tube should be installed, while a separate source of radiant heat should also be provided by an incandescent type lamp to allow the tortoises to bask in a dry position.

Housing Venomous Snakes

The majority of Australian snakes are venomous, and it would be impossible to produce a representative list of snakes suitable for keeping without listing some of these. Nevertheless persons under the age of 18 years or without adequate handling experience should not attempt to collect nor keep dangerously venomous species.

Collection Of Venomous Snakes

If you intend collecting any snake species, and are unfamiliar with the identification of all the snakes found in the area, it is imperative that you treat them all as potentially dangerous. If bitten by a potentially dangerous species, or one that you cannot identify with absolute confidence, you should immediately employ the pressure-immobilization first-aid treatment (see figure 9) and arrange transport to an appropriate medical centre.

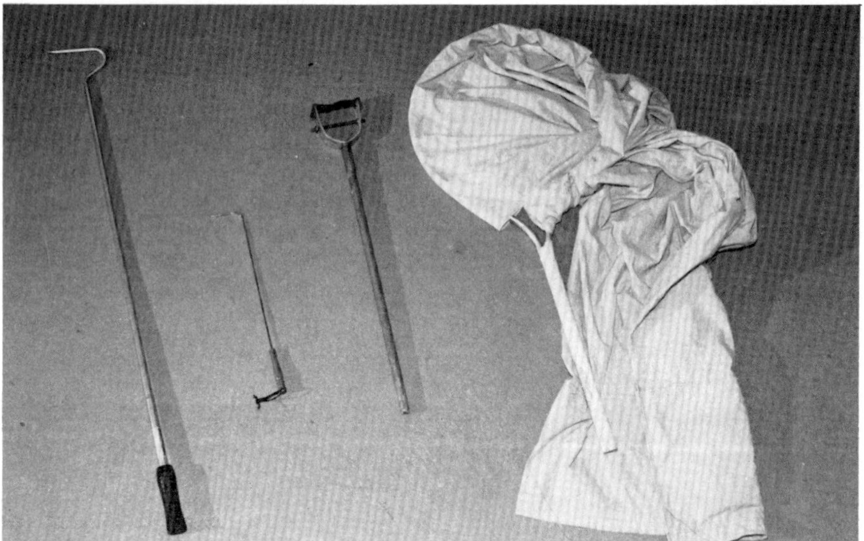

Figure 7 Useful tools in the handling of venomous snakes include snake hooks (*a*) and (*b*) for picking up and moving snakes, a 'jigger' (*c*) with suspended leather or rubber strap for pinning large specimens, and a hoop bag (*d*) to put snakes into. The snake hook and hoop bag should be the only tools used regularly in the day-to-day care of snakes, allowing the safe movement of venomous species into holding facilities.

Various tools, such as noosing poles, snake hooks, 'jiggers' and hoop bags can be used to safely capture potentially dangerous snake species. Contact your herpetological association for details as to the construction or commercial supply of these items, and if required arrange for a demonstration of the proper use of the tools.

Once captured, there is rarely any excuse for physically touching a dangerous snake with your hands or placing yourself or anyone else at risk of being bitten. The real experts never show off by handling their venomous snakes but rely upon the proper tools for moving their charges. They know that eventually, no matter how 'tame' a snake might seem to have become, it will one day be awakened in an irritable state, or perhaps confuse a moving hand or finger with food - in either case with potentially disastrous results.

Cage Security

Anyone keeping venomous snakes in residential areas has a responsibility and obligation to ensure against the possibility of the snakes escaping. This responsibility to recognize the safety of others is really no different to that placed upon owners of large dogs, guns, etc. In the case of captive, dangerously venomous snakes, standardization of cage construction has been established by the New South Wales National Parks and Wildlife Service. Diagrams and descriptions of several acceptable basic cage designs that are commonly used by reptile keepers and regarded as suitable for the housing of dangerous species are provided in figure 8. None of these cages are particularly expensive to construct.

Generally, adequate cage security depends upon a combination of sensible cage design, the use of suitable building materials, good workmanship, and the incorporation of an effective locking device. The net requirement is that the constructed facility must be strong enough to resist any normal destructive forces from within or without. Snakes are not known for an ability to dismantle cages; however, they are quite adept at discovering any and all weak corners, joints, gaps, or holes from which escape is possible.

Various sorts of locking devices requiring either keys or a numeric combination to unlock and allow the opening of the cage door are available and are easily installed. The purpose in locking dangerous snake cages is two-fold; not only to keep the snakes in, but to discourage unauthorized entry. No locking device will keep a determined thief or vandal from opening a cage, but at least family and visitors will be much less likely to open a cage if doing so necessitates destruction of the cage.

In the case where a cage door is not structurally rigid or is very large, all hinges and latching devices must be positioned in such a manner so as to minimize any 'play' in the doors. In other words, when the cage door is locked shut, there should be no possibility that it can be pushed outward at any corner to permit an escape.

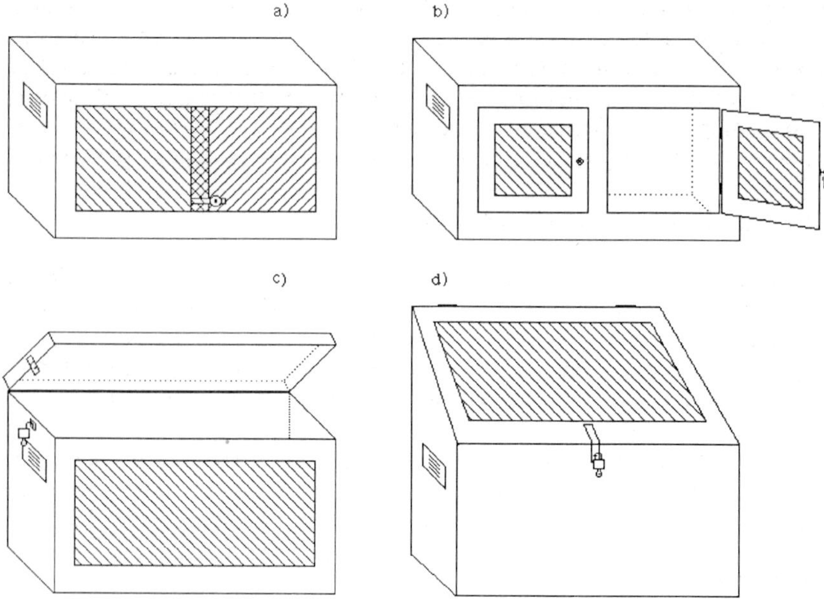

Figure 8 Cage security requires adequate attention to design and construction. Basic cage types include those with sliding glass doors (*a*), front-opening doors (*b*), and top-opening doors (*c*) and (*d*).

Cage Type A - Cages With Sliding Glass Doors

Sliding glass doors can be locked with shop display (ratchet style) locks, or similar locking devices that require either a key or combination to open. Beware of gaps greater than 3mm (or less for juveniles) between or above doors. If need be, glue a strip of glass or perspex to the overlapping edge of either door in such a manner as to occupy the excess gap between the two doors when closed. Many commercial glaziers (such as O'Brien's Glass) offer reasonable package prices to install sliding glass doors, complete with narrow plastic grooves, ratchet style locks and handles or finger grooves. Always have the edges of glass doors smoothed for ease of sliding and to avoid damaging snakes' snouts.

Cage Type B - Cages With Swinging Hinged Doors

Cages with swinging doors can be locked with standard cabinet locks, hasps or clasps - with padlocks, or any other locking device that adequately locks the door(s) shut by means of key or combination. If twin doors are to be locked to one another, make sure that at least one door is also locked to cage structure. Make sure that no gaps greater than 3mm (or less for juvenile snakes) exist around the edges of all doors when closed. In the case where a door is not structurally rigid, be sure to position hinges and latching devices in such a manner as to minimize any 'play' in the doors (particularly at lower corners) when pressure is applied from the inside.

Cage Type C - Lid/Top Opening Cages

Cages that open from above can be locked with padlocks (in conjunction with hasps), or any other device which can be used to lock door shut by key or combination. In the case where a door is not structurally rigid, be sure to position hinges and latching devices in such a manner as to minimize any 'play' in the doors (particularly at front corners) when pressure is applied from the inside.

Top-opening cages can be awkward or even dangerous to work with if they are not fitted with a means of holding the lid open. You may need to install a hinged brace, or position the cage so that when raised, the top rests against the wall behind it.

Escape-Proof Air Vents

Fly-screen, wire mesh, or prefabricated products can be safely used as air vents in snake cages, providing they are properly affixed. Great care should be taken if fly-screen is used, and it is recommended that only small vents be fitted. To test the strength of an air vent, push it at all points from the cage's interior with a single finger. Use the same amount of force that you would use to register a reading of 3 kilograms on a balance. If possible, first test yourself by using the same finger to press on a set of bathroom scales. The vent should be able to withstand this amount of pressure anywhere.

Security of the Room

To ensure the security of any snake keeping facility, important consideration must be given to the actual room in which the snake cages are to be held. An escape-proof room offers the additional assurance that even in the event of a cage being inadvertently left open, any escapees can be reclaimed. It also helps to ensure good relations between you and your neighbors, some of whom may not be kindly disposed to snakes. Making a room 'snake'-proof' is generally a fairly simple procedure and well worth the time and minor expense involved, considering the added security it affords. The most serious consideration in most cases, is simply the elimination of any gaps that may exist around the room's door(s) when closed. This can be done cheaply and effectively by affixing a narrow strip of timber to either the bottom of the door or the floor. There should not be any space around the door when it is closed. All windows in the room should be kept locked or properly fitted with quality fly-screens, preferably double fly screens. There should be no gaps between the fly-screen and the window frame. Check for holes in the floor, walls, and ceiling, including closets, cupboards, etc. Look for air-vents, disturbed telephone or electrical connections, etc., and if any gaps exist, seal them.

First-Aid Treatment for Venomous Snake Bites

If you adhere to the snake caging requirements discussed above, use the correct tools and exercise appropriate caution in the maintenance of your venomous snakes, you will not be bitten. However, anyone who keeps venomous snakes should be well versed in the application of the proper first-aid treatment for snake bites. The technique overwhelmingly recognized as the most effective in slowing the effects of envenomation is the 'pressure-immobilization' technique. This is based on the fairly recent discovery that the venoms of Australian snakes initially spread through the body from the bitten region by moving through the lymphatic system, not the blood stream. The pressure-immobilization technique acts to slow or nearly stop the flow of the lymphatic fluid away from the bite site. This is relatively easily accomplished since the lymph vessels are positioned just under the skin, so a tourniquet is not required. If properly applied, the first-aid will halt the progress of envenomation long enough for the victim to arrange transport to hospital for medical treatment which might include the use of anti-venom. Bite victims should try to relax and reduce blood flow. There is no need for panic; they are not going to drop dead instantly (and it is extremely unlikely that they are in mortal danger).

If you keep venomous snakes, you should have several crepe bandages on hand, and if you collect snakes from the wild, you should also always carry one or two of these bandages with you. If someone has been bitten and for some reason there are no bandages on hand, improvise with strips of clothing, stockings, belts, etc. It would be prudent to telephone ahead to the hospital so that they can get prepared in advance.

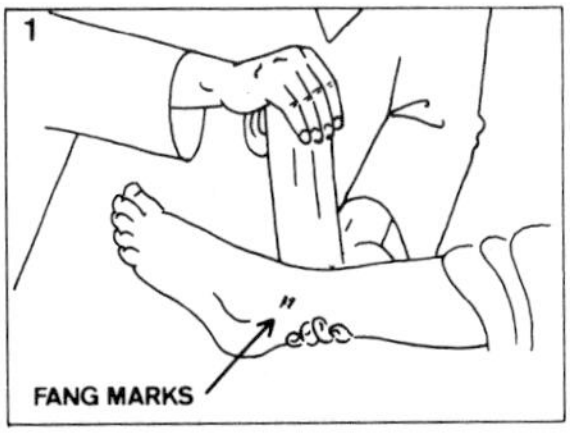

1. Apply a broad pressure bandage over the bite site as soon as possible (don't take off jeans as the movement of doing so will assist venom to enter the blood stream. Keep the bitten leg still!).

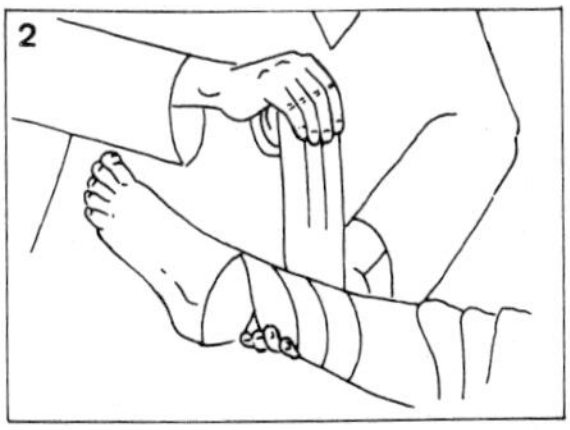

2. The bandage should be as tight as you would apply to a sprained ankle.

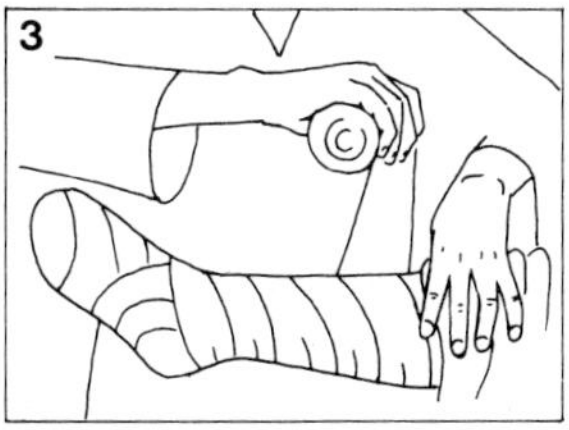

3. Extend the bandages as high as possible.

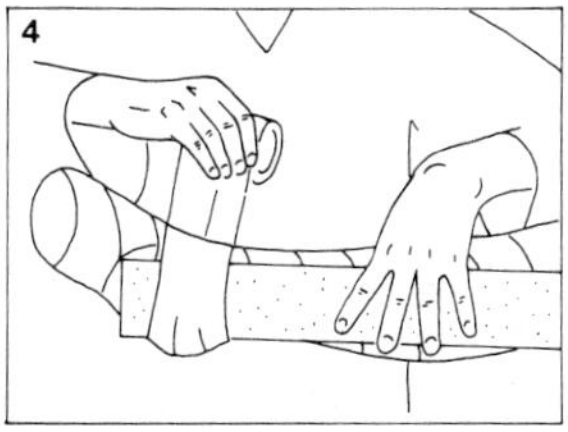

4. Apply a splint to the leg.

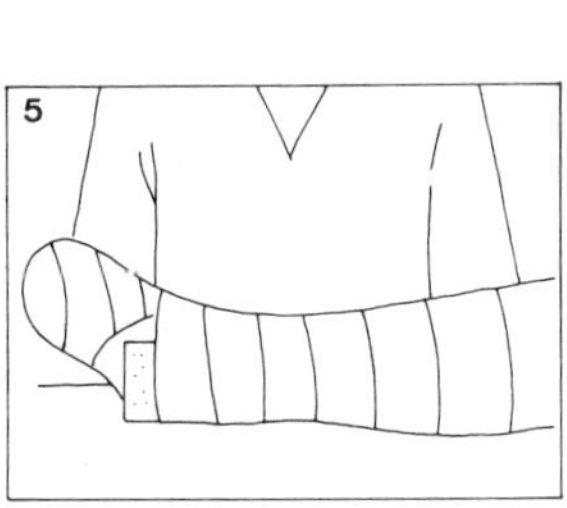

5. Bind it firmly to as much of the leg as possible.

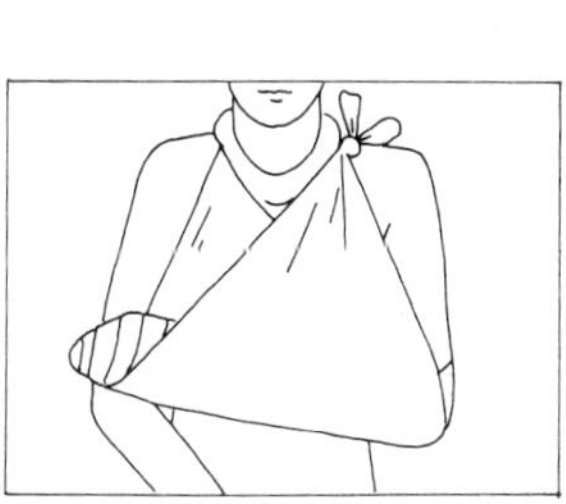

Bites on hand or forearm. **1.** Bind to elbow with bandages. **2.** Use splint to elbow. **3.** Use sling.

Figure 9 Diagrammatic representation of first-aid treatment for snake-bite victims (after Dr Struan Sutherland of the Commonwealth Serum Laboratories, [1979], *Family Guide to Dangerous Animals and Plants of Australia*). In his description of this technique, Dr. Sutherland writes: "To apply pressure, crepe bandages are ideal, but any flexible material can be used, for example, torn strips of clothing or old towels. Bandaging should be over the bitten area first and then extended from the toes or hand upwards. Although this may squeeze some venom towards the body, it effectively empties vessels as it compresses and the end result is probably more comfortable. If the patient complains strongly of discomfort in the bandaged limb then it has been too tightly bound and should be modified. It is very important that, if the leg has been bitten, walking should not be attempted, as this will assist venom movement"

Feeding

A reptile's willingness to feed in captivity is in part an indication of its acceptance of the circumstances of its confinement. Before you can expect your reptile to feed it must become familiar with and feel secure in the environment that is provided. To ensure this occurs you must take into account the sorts of caging and husbandry considerations discussed in this book. A settling in period of up to two weeks may be needed by some freshly captured specimens before feeding commences - possibly longer for the larger snakes.

Always maintain detailed feeding records of your reptiles so that you can be sure that each specimen is being fed adequately. A well kept record book will make it possible to identify any unfavourable trends in feeding behaviour. The information that should be recorded are feeding dates, what was offered, and what was accepted. Along with your feeding records, it is also a good idea to record any other significant events as they occur (e.g. dates of skin-sloughing, dates and descriptions of any medical treatment, dates of observed mating, etc).

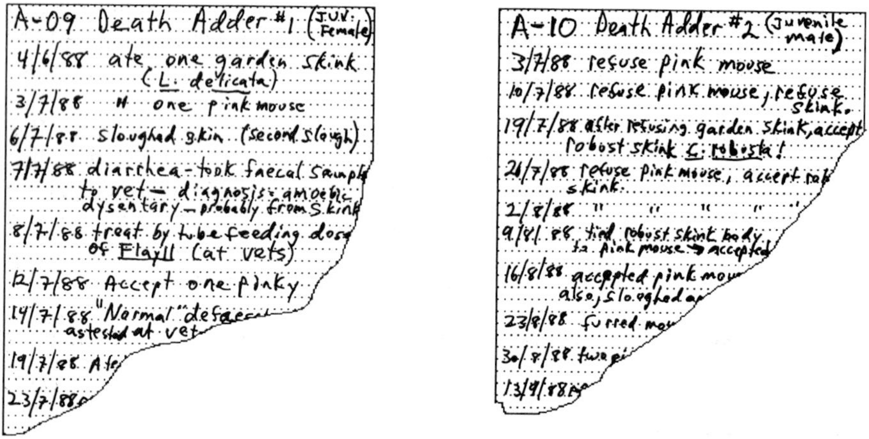
A-09 Death Adder #1 (JUV. Female)
4/6/88 ate one garden skink (L. delicata)
3/7/88 " one pink mouse
6/7/88 sloughed skin (second slough)
7/7/88 diarrhea - took faecal sample to vet - diagnosis: amoebic dysentary - probably from skink
8/7/88 treat by tube feeding dose of Flagyl (cat vets)
12/7/88 Accept one pinky
14/7/88 "Normal" defaec... as tested at vet
19/7/88
23/7/88

A-10 Death Adder #2 (Juvenile male)
3/7/88 refuse pink mouse
10/7/88 refuse pink mouse, refuse skink.
19/7/88 after refusing garden skink, accept robust skink L. robusta!
26/7/88 refuse pink mouse, accept robust skink.
2/8/88 " " " "
9/8/88 tied robust skink body to pink mouse → accepted
16/8/88 accepted pink mouse also, sloughed
23/8/88 furred mouse
30/8/88
13/9/88

Figure 10 A record booklet should be maintained, indicating dates of feedings, medical treatments, sloughing, matings, length and weight changes and other particulars for each specimen being kept.

The natural diets of reptiles differ considerably. The types and quantities of foods taken by a reptile in its natural environment will obviously reflect the nutritional needs of the species. Therefore you should strive to provide your charges with the sorts of foods they would naturally feed upon. This might involve the breeding or collecting of anything from mealworms and grasshoppers to frogs and small lizards (subject to legal restrictions). The problems associated with feeding tortoises, snakes and lizards are fairly different, so they are dealt with here separately. Following these discussions, is a section devoted to feeding problems that may arise, what to do if a lizard or snake refuses to feed, and information on vitamin and mineral supplementation.

Feeding Tortoises

Although tortoises are mainly flesh eaters, some (including the Macquarie tortoise and other 'short-neck' species) will eat water plants and other forms of vegetation occasionally. Adult long-necked tortoises do not eat substantial amounts of vegetation in the wild and will generally refuse it in captivity. Tortoises normally prefer to feed while in the water, but will soon learn to recognize you as the source of food and become trained to come out of the water and take food from your fingers. Suitable foods include large insects, earthworms, whole raw fish, yabbies, ground beef, etc. An all-meat diet is quite deficient in vitamins and minerals, and will need to be supplemented with a vitamin mineral mix (see section on vitamin and mineral supplementation) or replaced with a more varied diet. Some tortoises will readily accept commercial fish food pellets which have essential vitamins and minerals.

Hatchling tortoises need to have their food chopped into tiny pieces, or fed small live insects and crustacea. Tortoises kept indoors may be moved into a separate tank to be fed, since they will otherwise dirty their home environment during feeding. Care must be taken to avoid leaving any uneaten food in either tank. Hatchling tortoises require special attention. Various water plants should be established in the vivarium (ultraviolet lighting will help promote plant growth). With these plants will be found small invertebrates upon which the tortoises will feed. However, more small insects should be added regularly. These can be collected by dragging a net through a pond or fresh-water creek.

Feeding Snakes

All snakes are predators and most kill and eat other vertebrate animals in the wild state. Prey items are either suffocated by constriction (pythons and some elapid snakes), killed by venom or swallowed alive (green tree snakes swallow live frogs).

In any event, all swallow their prey whole, and are able to manage rather sizeable meals - due in part to the extreme elasticity of the entire digestive tract, and the fact that the skull and lower jaws are modified for the task. Specific feeding requirements for particular species are discussed in the care sheets provided for them.

Snakes normally only eat whole animals, from which they are able to obtain any necessary vitamins and minerals. They shouldn't require any sort of vitamin supplementation in captivity if fed the appropriate types of food, though it is possible that they may benefit from exposure to ultraviolet light. From a nutritional standpoint, it probably makes little difference what sort of vertebrate animals are eaten, however, most species are fairly specific about what they will accept.

Obtaining Snake Food

Before deciding what species of snake you would like to keep, you should carefully consider the availability of suitable food for it. Species that normally eat frogs or lizards may or may not be willing to make the switch to a more easily supplied diet of laboratory rodents. If you choose to keep a snake that requires either frogs or lizards as food (and most do at least as juveniles), you may be committing yourself to spending a good deal of time and effort wallowing about in local swamps and irrigation ditches for frogs or perhaps chasing small lizards in nearby paddocks.

There are a number of protected frog species that cannot be collected for snake food, and only a few lizard species are categorized as 'exempt' for use as reptile food for NSW hobbyists.

It is generally a much simpler task to provide food for snakes that accept warm-blooded prey. Chickens, from day old hatchlings to full-sized adults are often available from commercial hatcheries, and laboratory mice and rats can either be cultured by yourself, starting out with a few adult rodents or purchased from a local pet shop.

Weaning Predominantly Frog or Lizard-eating Snakes onto a Diet of Warm-blooded Prey

Luckily, with a bit of persistence you can often take advantage of the importance of scent detection by your snake in the identification of its food. By rubbing the scent of the snake's preferred food (eg a frog or skink) onto say, a dead mouse, which wouldn't normally be accepted, you may be able to trick your snake into eating unfamiliar food items. If the mouse is eaten, repeat the same operation during each of the next few feedings - perhaps using a bit less frog scent each time, until eventually the smell of the mouse itself is enough to stimulate a feeding response. If small lizards are to be used as the source of scent enticement, the dead lizard can be slit along its belly, and the intended food item coated with its blood and internal fluids.

Alternatively, if the scenting trick described above doesn't achieve any success, you can tie the preferred food item (eg dead frog or skink) onto the unfamiliar one with a strand of plain white cotton thread. It is important to use as little cotton as possible, with no loose ends. Some success may be had in inserting the broken tail tip of a feed lizard into the mouth of the dead rodent.

Besides the advantage of the easy availability of laboratory rodents, in weaning your snakes away from frogs and lizards you reduce the risk of introducing internal parasites to your snakes. If these food items must be used it is advisable to freeze them beforehand as a humane way of killing them and in order to kill at least some of the parasites that they harbour. Keep them frozen for at least 60 hours, but make sure any frozen food is fully thawed before being offered to your reptiles.

Separating Snakes Prior to Feeding

Once your captives are accustomed to whatever feeding procedure you adopt, they will become much more at ease with the operation, and very easy to feed. However, where two or more snakes are normally kept together in the same cage, they should be separated prior to feeding, otherwise they risk the possibility of injuries or accidental cannibalism if more than one of the snakes attempts to eat the same food item. You may also find that a snake will refuse to eat in the presence of another. If you do not have extra cages for feeding purposes, plastic Chinese food containers or equivalent with tiny air holes punched in the lids can be used for feeding small individuals; either large cloth bags or plastic rubbish bins with fastening lids can be used for larger snakes. If you are planning to keep a reasonable number of snakes, it is a good idea to invest in a couple of these bins, as they are also very handy for holding the snakes while cleaning their cages. Of course, any rubbish bin to be used to hold or feed animals should never be used for its original purpose. Always move the snakes that feed readily into the temporary feeding quarters, leaving the shyest specimen to feed in its accustomed surroundings.

Will Snakes Accept Dead Prey?

The majority of captive snakes will quickly adapt to a diet of dead (even defrosted) prey items providing that these are of the right sort. The first few meals of dead prey might involve some coaxing. Perhaps the best trick is to wriggle the food item with a long pair of forceps in front of the snake as if it were still alive. If the snake grasps the prey item, fine; if it doesn't, leave the prey in the cage for a few hours and perhaps in your absence it will be eaten. NEVER leave a live adult rodent unattended in a cage with a snake. Apart from being cruel to the intended prey, if the snake is not interested in feeding, the mouse or rat may become frightened or hungry and eventually attack and injure the snake. In confined circumstances, snakes sometimes fail to respond defensively, and even highly venomous species may be hesitant to bite the offensive rodent.

How Big Should a Snake's Meal Be?

Snakes are able to swallow surprisingly large prey items, which is one of the reasons that they are able to feed so infrequently. A prey item (or combination of prey items) with an approximate weight of, say, 5-20% of the snake's weight would be considered a reasonable meal. Perhaps a better way of putting it is that a sensible meal would create a slight bulge in the snake's midsection.

How Often Should Snakes be Fed?

Many snake keepers adopt a once a week feeding schedule - designating a particular day of the week when all of the snakes are fed. Those that are unwilling to feed or are in the 'opaque' stage of shedding the skin simply skip that meal. Certainly a once weekly feeding schedule will more than suffice if the meals are of adequate size, and in fact fortnightly feedings would probably be enough for less active species such as most pythons. In the event that your snake doesn't want to feed for a while - don't panic: remember that a healthy snake that is not subjected to undue stress will typically be able to go without eating for months at a time without any substantial loss in body weight. In the interest of breeding, some species are best cooled down during winter months, during which time feeding should be discontinued. This subject is further discussed in the reptile breeding section.

As a rule, captive snakes tend to be overfed. One needs only to compare a typical long-term captive with a freshly caught wild specimen of the same species and of equal length to see the difference in body proportions! In the case of captive-raised snakes, overfeeding typically leads to an excessive body girth, with a smaller than normal head (possibly the result of the snake growing too fast for the correct development of the skull bones). It is believed that obese snakes live shortened life spans. Of course, the other extreme is also to be avoided; underfed young snakes develop into stunted adult specimens.

When to Feed Snakes

The activity periods for snakes vary considerably. You will probably notice that your snake is most active at particular times of the day, during which it will be most likely to be interested in feeding. Diurnal sun lovers are best fed during the day, and are unlikely to be interested in eating at night. Snakes that are nocturnal and would normally feed at night may need to be offered food either in darkness or in simulated darkness with the use of a coloured light source (see lighting section). Following feeding, snakes should be kept warm and not handled or otherwise disturbed.

Feeding Lizards

Lizards are quite varied in their food preferences, though most species are at least partially predatory, with insects and other small invertebrates comprising the greatest part of their natural diets. Specific feeding requirements for particular species are discussed in the care sheets provided for them.

Supplying Insectivorous Lizards With Food

Insectivorous lizards should be fed a fairly wide variety of insects if you are to ensure their good health in captivity. Insects and other invertebrates feed on a tremendous variety of foods which in turn varies their chemical makeup and nutritional value as lizard food.

The techniques for raising mealworms, locusts, crickets and fruit flies are outlined below. These insects can form a good dietary foundation for most insectivorous reptiles, but when time and opportunity permit, it is a good idea to occasionally supplement these cultured insects with a selection of wild caught ones. Various techniques exist to trap moths and other nocturnal insects, and with the use of a simple sweep net you can catch a great variety of insects, especially grasshoppers. Beware of insects that may have been subjected to insecticides as these chemicals will be passed on and accumulate to lethal levels in the lizards that eat them.

Culturing Mealworms

Mealworms are an easily reared food source for many small lizards and their cultivation is rather straightforward. However, they are generally considered to be sub-optimal in terms of food value and relatively difficult to digest. The excessive amount of chitin (the hard material that makes up the exoskeleton of insects) present in mealworms, both in larvae and adult beetles, has been known to cause intestinal blockages in small lizards fed exclusively on them. It is best to use mealworms which have recently 'shed' their hard, chitinous outer layer and before their new exoskeleton has hardened. At this stage they are white in colour and are much easier for lizards to digest. It is wise to use mealworms sparingly, more as a supplement than as a foundation for your reptile's diet.

Most successful mealworm colonies seem to be kept on a single-container principle: all growth stages (mealworms, pupae and beetles) are maintained together for both breeding and continuous collection of large mealworms for feeding. Additional food is added to the colony from time to time, and it is usually necessary to clean out the container every few months.

Any large containers with steep, smooth walls, such as plastic ice cream containers, can be used to house the colony. The container should be filled to within two or three centimetres of the top with a mixture of two parts (by volume) of bran, to one of pollard. A few tablespoons of wheat germ can be added. Always leave the lid off the container, or use a well ventilated lid if escape by the insects is possible, otherwise condensation will create a problem. If the bran/pollard mixture becomes wet, it will become a sticky mess and will lead to the decline of your colony, so the food should be replaced.

Next, add a single layer of wide strips of newspaper on top of the bran and pollard mixture. Leave a gap between the newspaper and edge of the container to prevent beetles using it as a point to climb the sides. Place occasional bits of fruit or banana peel on top of the layer of newspaper as a source of water for the insects.

Temperature of the colony should be maintained between 27° and 30°C, and the relative humidity between 40 and 70% RH. The latter humidity range should result if water is sprinkled (sparingly) over the newspaper every couple of days. Addition of water will also cause the mealworms to migrate from the medium to the paper to drink, and thus facilitate harvesting. The mealworm colony should be maintained on a cycle of 12 hours light, 12 hours dark.

To get your culture started, you may be able to obtain mealworms from one of your fellow reptile keepers. Otherwise, contact your herpetological association or a local fauna park for supplies. It generally takes about two months to build a sizeable population from a dozen or so initial beetles. The beetles begin laying five to ten days after hatching. Each female lays 350 to 400 eggs during her life cycle. Fertility drops sharply after about four weeks,and at this age beetles are best used for reptile chow. The eggs hatch in eight to nine days and remain as larvae (mealworms) for nine to twelve weeks.

Growth of mealworms can safely be arrested for at least six weeks by placing a small container of them in your refrigerator. This container should be half filled with dry bran and covered with plastic to retard dehydration. The 'hibernating' mealworms should be taken out once a fortnight or so for a full day in order to be fed with carrot.

Culturing Locusts

Locusts are a good food source for many reptiles - and are about half way between mealworms and crickets in terms of food value. There are many species, with the best ones breeding throughout the year and producing large numbers of eggs per clutch, and many clutches per year. To obtain your starter colony, you will need to contact your fellow reptile keepers, your herpetological association, or any nearby fauna parks. Otherwise contact the department of agriculture in your State government for advice on obtaining initial breeders or for pointers on culturing them.

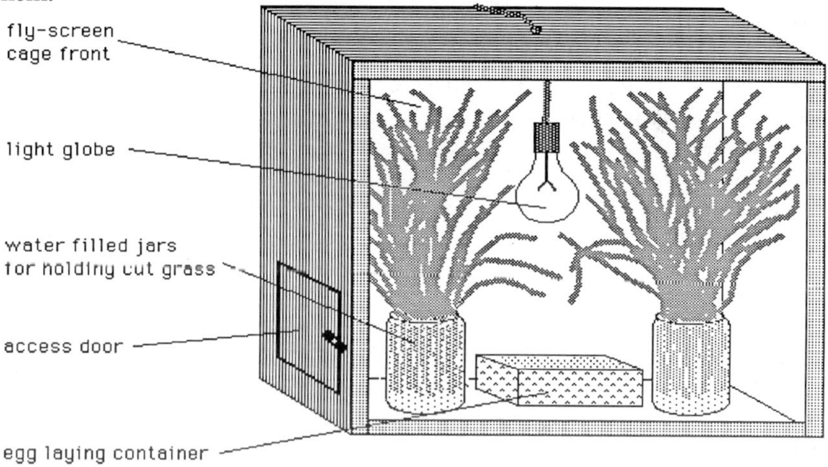

Figure 11 Locust breeding cages should be fairly tall to accommodate grass cuttings and large numbers of locusts, which prefer to position themselves upon the walls and screen front.

At least two cages are needed for breeding locusts: one for the adults to breed in, the other for the eggs to be transferred to and the young to be raised to food size. Both cages should be fairly spacious - 40cm x 40cm x 60cm is minimal, with glass, perspex or screen fronts and(or) walls. Fly-screen sides have the advantage of allowing climbing by the locusts, but result in excessive heat loss. Access to the cage can be provided by a hinged door in the front or side that should be just big enough to allow your arm to reach in comfortably (approximately 20 cm x 20cm).

The temperature maintained in the cages is critical. It must be maintained at approximately 32°C (during the day) or the colony will slowly deteriorate. This high temperature is best provided by hanging an incandescent light globe fixture in the centre of the cage. You can establish a light/dark cycle, or just leave the light on continuously 24 hours per day for faster development. Humidity will not be a problem if adequate ventilation is provided.

Locusts must be fed daily with fresh cut grass (broad-leaved types are best). Make sure it has not been exposed to insecticides or fertilisers. The grass should always be available to the insects, particularly hatchlings, which can starve in only 7 hours. Also, locusts turn to cannibalism in the absence of appropriate food. The best way of offering the grass is by placing it tightly in a glass or small jar filled with water in the same way that flowers are displayed in a vase. That way the grass won't dry out before being eaten. Make sure that the grasshoppers cannot fall into the water and drown. Cages must be cleaned regularly and uneaten grass, as well as droppings and dead insects should be removed.

Twenty-five pairs of adult locusts may be kept in a cage with a floor area of 60cm x 40cm and a height of 60cm. The females will lay 30 to 100 eggs per clutch. You must provide containers with a slightly dampened sand and soil mixture for the gravid locusts to lay their eggs in. Rectangular 'Tupperware' products with walls of 10 to 12 cm height are ideal. Fill the containers with a sand/soil mixture and place one of these either on the cage floor or, preferably, sunken into an appropriately sized hole in the cage floor - so that the surface of the egg laying medium is flush with the cage's (false) floor. The reason for doing this is that locusts are more likely to lay in the container if it is flush with the ground. If the container is too high, a sizeable proportion of clutches will be laid on the cage floor and wasted. Once a week these containers need to be changed over with one of two that will be kept in the nursery cage. Label the containers with a date as they go into the nursery to ensure that no egg containers are removed prior to having been in the nursery for a full two weeks. You will probably have to sprinkle some water on the hatching medium or soil from time to time to keep it slightly moistened.

The eggs take 10 to 14 days to hatch and the young, which will be only 1 cm long, immediately feed on grass (which should be available at all times in the nursery cage). The hatchlings will grow through five developmental stages, reaching the adult stage in about five to six weeks. Within a few weeks of reaching adulthood, they will be able to reproduce.

You should continuously replace your breeders with raised young since adults become infertile after a year or so, depending upon the species. Most reptile keepers find that the best locusts for feeding to reptiles are those harvested while in the two developmental stages just prior to the adult stage, since these ones are softer bodied (and therefore more digestible) and cannot fly. If you find that you have large numbers of locusts at the prefered stage of development all at once, freeze some of the excess for later use. Most locust-eating reptiles will relish thawed locusts offered on the end of a pair of forceps.

Culturing Crickets

Various species of ground dwelling crickets can be cultured to provide the dietary foundation of many insectivorous reptiles. They are extremely high in food value and easily digested. Cages for breeding them should have extensive floor space. A good size for a cricket breeding cage would be 70 cm wide, 40 cm deep and 40 cm high. The materials used for construction are not particularly important, though a transparent front allows easy viewing. The cages can be roofed with a hinged wooden frame covered with fly-screen. The floor should be covered by a 5-10 cm layer of peat moss, over which are scattered a few pieces (30 cm x 30 cm) of cork or similar material, beneath which the crickets will hide by day.

Watertight containers can be fitted with moistened sponges to provide drinking water. Any standing water left in the cricket cage will merely provide a drowning pool since the crickets will fall in and will probably not be able to crawl out again. As soon as the peat floor covering gets grainy, the cage should be completely cleared, the peat discarded (or temporarily held in another container for two weeks to hatch any eggs deposited within) and the cage cleaned and returned to operation.

Crickets must be kept fairly warm - 30°C seems to be ideal. Since they will need a 12 hour day, 12 hour night cycle - or thereabouts, you might consider heating the cage at night with a red or blue light globe to simulate darkness while still providing sufficient warmth.

Fresh lettuce leaves, cabbage leaves, cut pieces of carrot, apple or other vegetable or fruit should be added daily to the peat, and any left-overs should be removed. A small container full of dry oat flakes or dry dog food should also be kept in the cage.

Each breeding cage of the above size can be stocked with approximately 50 pairs of adults. The females use their ovipositors to place the eggs directly into the peat layer. The young hatch after about 14 days and should be left in the breeding cage with the adults. The young pass through five developmental stages to become adults after six weeks. Within a couple of weeks these new adults will also be laying eggs.

Once your colony(ies) of crickets become established, you will have crickets of all stages of development on hand at all times, and will be able to cull excess stock for feeding purposes as required. The easiest method of catching them is to lay some wide-mouthed bottles or the cardboard tubes from the centres of toilet rolls on the peat surface. Almost all of the crickets can be caught within a few days using this method. The largest individuals can be cropped to restock a new breeding cage. It is recommended that several colonies be maintained simultaneously if you have many insectivorous reptiles to feed.

Culturing Fruit Flies

Fruit flies are the very tiny flies that are frequently detected buzzing around rotten fruit. They can be easily cultured to feed the tiniest of hatchling, insectivorous lizards. Initial 'breeders' may be lured with rotten fruit, or can be obtained from the biology laboratories of most universities or colleges. You may be able to obtain genetically mutant stock that is wingless - a very practical feed insect.

To culture fruit flies, simply take a large glass jar and add a layer of crumbled stale bread over the bottom to a depth of about 3 cm. On top of this dry layer, add ripened fruit - bananas or apples will work well. As the fruit deteriorates, its juices will be absorbed by the bread crumbs below. A little dried yeast is then mixed in a small quantity of water and poured over the fruit to hasten fermentation. A few fruit flies are then added and the jar covered with a piece of cotton cloth secured with a rubber band. The jar should be placed in a position that will maintain a temperature of between 20° and 25°C.

Feeding Canned Pet Food to Lizards

Many lizards will accept canned dog or cat food as a dietary staple - in fact some species (eg pink-tongued skinks, common bluetongues, and blotched bluetongues) can be expected to live healthy, active and breeding lives on a sole diet of this food. There are a number of suitable makes on the market - though it is important to avoid using ones high in fats or fish products, or ones over-supplemented with vitamins.

How Much Should Lizards Eat?

Lizards, as a rule are more active and have a higher rate of metabolism than snakes. They generally feed far more frequently than snakes - though admittedly in smaller 'bites' and individual meal sizes. Small lizards can lose weight and general body condition very quickly if not provided with adequate nourishment. It is vital that your captive lizards are fed on a regular, scheduled basis, and are watched individually for signs of failing health or weight loss. Most small insectivorous lizards should be fed every couple of days - the size of each meal being perhaps half the total quantity of food that they are capable of consuming. The larger species can for the most part be put on a once-weekly feeding schedule. The important thing is to establish a feeding schedule and stick to it.

When to Feed Lizards

Rather than simply depositing large quantities of food in a lizard cage and replacing it as it disappears, it is better to feed the lizards individually (with a long pair of forceps if need be) and at specific times so that you can keep accurate records of who ate what and when. The best time to feed lizards is during the hours when they are most active. Therefore, the diurnal ones should be fed during daylight hours-preferably at a time when they appear to be 'on the prowl', while the nocturnal ones may be fed during the night phase of their vivarium environment, though they will probably adapt to feeding by day.

Feeding Problems

If you find that a lizard or snake is simply not interested in eating, and shows signs of deteriorating body condition, you will need to identify the reason(s) for the unwillingness to feed. But first you must be able to recognize a 'deterioration in body condition', which in itself can be most difficult for the novice reptile keeper. Snakes with low body weight may exhibit unnatural skin folds, or jutting vertebrae or ribs. Lizards may exhibit similar signs when underweight, but many species (eg. geckos) quite naturally have protruding skeletal features regardless of relative body weight. A sure sign of a starving (or dehydrated) skink, goanna or gecko is an emaciated tail and prominent hip bones. The tail for these lizards (in some species more than in others) acts as a fat store, and will decrease in girth as these reserves are used up. Underfed tortoises will become noticeably less heavy, and the fleshy areas surrounding the limbs, tail, and neck will show signs of receding further into the shell. If you are not sure your reptile requires attention, contact a more experienced reptile keeper for assistance or advice.

Reptiles that steadfastly refuse to feed can be divided into two distinct groups: those that have only recently been born/hatched or captured, and those that have suddenly stoped eating after having apparently adjusted to captivity.

New Arrivals that Won't Feed

You can expect newly born, hatched, or freshly captured specimens to be reluctant to feed for the first few weeks of captivity. New born reptiles in particular, often do not feed for a few weeks after birth, their food needs being supplied by residual yolk. Until a reptile has 'settled in', it should have its own cage, and of course be disturbed as little as possible. If a new captive is subjected to excessive handling, provided with unsuitable hiding facilities, or subjected to conditions that are too cold or warm, you could be waiting indefinitely for it to commence feeding. A logical question to ask yourself when assessing the new arrival's cage facility is: Have I provided as closely as possible and practical, a captive scenario that reflects the natural environment of the reptile, particularly in relation to climate and substrate choice? Make sure that the reptile has found and is using the hide box or other retreat provided, and that the temperature within the retreat is appropriate.

A second step to take in your attempt to determine the cause of the new arrival's lack of appetite is to review the way in which the food has been offered. In the case of snakes, you should not feed until the snake settles into a hide box or similar retreat. Then place a dead prey item in the entrance for a couple of hours. If practical, you might cover the cage with a cloth sheet to avoid distracting the snake with your activities. If the prey item is not eaten, try a live one presented in the same manner. You can put a frog or lizard in the hide box with the snake, but when feeding rodents, leave the live prey outside the retreat. As mentioned previously, do not leave a live adult rodent unattended with any snake. Lizards can be offered their food item on the end of a long pair of forceps. You can also try pressing the food item right up against the lizard's snout.

Every potential prey species has its own unique odour. You may find that a young snake that steadfastly refuses to take a 'pink' mouse will eat a pink rat, or perhaps a mouse with just a bit more size - and hopefully a more enticing scent or appearance. If you are finding that a frog-eating species, such as a swamp snake is refusing the frogs you offer it, try a different type of frog - or if allowed in your state or territory, perhaps a designated species of 'feed lizard'. Just because an insectivorous lizard refuses to take one type of insect doesn't necessarily mean it won't relish another - don't give up until you've tried grasshoppers, crickets, cockroaches, and moths. When all else fails, mix an uncooked egg in a shallow dish and put the lizard's snout in the egg. It's surprising how many different species relish this treat.

Long Term Captives that Won't Feed

If a lizard or snake that is a long term captive has not fed for a long time, after previously having fed well on a regular basis, the first consideration must be to assess its health in an attempt to rule out any obvious disease or parasitic infestation. If you are able to procure a faecal sample, get your veterinarian to inspect it microscopically. Check the reptile carefully for external parasites (mites or ticks) and also check the shelter and cage for such parasites. Also inspect the reptile's mouth for signs of pneumonia or canker (see disease section of this booklet). Is the skin in good condition? Is the reptile's body temperature being maintained high enough? Failing this, consider whether or not the reptile is overweight; often an obese specimen will stop feeding until its body weight becomes somewhat reduced. This can take months for snakes, in which case patience will be your only recourse.

You will find that reptiles will sometimes feed and grow in spurts, and a healthy reptile will not necessarily feed on every occasion that you think it should. Many species, particularly pythons, will not feed at all during the winter months, even if the cage temperature is not substantially lower than the level at which it is kept during the warmer months of the year. This is perfectly normal behavior, reflecting the seasonal fasting that occurs in the wild.

Involuntary Feeding of Lizards and Snakes

The involuntary forced feeding of snakes and lizards is stressful and should only be considered as a 'last option' when a specimen steadfastly refuses to feed voluntarily while its health deteriorates from lack of nourishment. Dangerously venomous snakes should only be involuntarily fed by a highly experienced person, and only then in exceptional circumstances. Sometimes involuntary feeding can trigger a reptile into accepting additional food on its own, or at least keep it sustained until the cause of its lack of appetite can be ascertained. There are records of snakes being maintained by this method for years, but when your reptiles are not feeding for themselves, you can assume that something is wrong - either with their health, or their captive environment.

Involuntary feeding includes several distinct techniques of administration. Those discussed here are: 'assist-feeding' - in which food items are only introduced into the reptile's mouth with the hope that the feeding response will follow; 'force-feeding', where an entire food item is forced down the oesophagus of the reptile towards its stomach; and 'tube-feeding' whereby a mixture is introduced directly into the reptile's stomach via a long, flexible tube. If you are contemplating any form of involuntary feeding but have not had any previous experience, you should get help from someone who has, or seek the assistance of your local veterinarian. The risk of injury to the fragile mouth parts and tissues is great if the techniques are applied incorrectly.

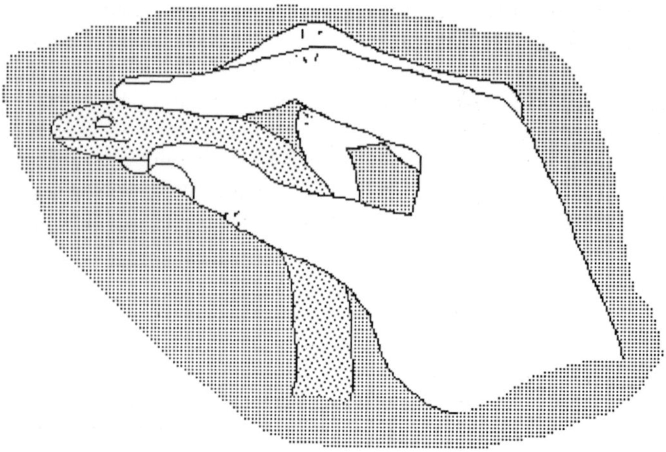

Figure 12 If a venomous snake must be held for force-feeding or treatment with medication, it may be pinned immediately behind the head with a snake jigger on a soft substrate such as a foam pad or carpet and carefully grasped with the forefinger on the top of the snake's head and the thumb and third finger just behind the angle of the snake's jaws, leaving the other hand free to support the body. This technique should be practiced extensively with non-venomous species before attempted with venomous ones.

Assist -feeding

'Assist-feeding' refers to the forced introduction of a food item (usually a dead prey animal) into the fasting reptile's mouth, leaving it to be swallowed voluntarily. This is generally a much less stressful exercise than actual 'force-feeding', where the food item is forced down towards the reptile's stomach. Additionally, when successful, assist-feeding is more likely to elicit a resumption of normal feeding behaviour. To assist-feed a lizard or snake, you will first have to open its mouth. This can usually be accomplished by using the food animal's snout or head to gently pry the mouth open, being very careful not to scratch the inside of the reptile's mouth. To assist-feed skinks, mild pressure on both sides of the tip of the snout will often induce opening. In the case of bluetongues, any contact of food within the mouth will cause the lizard to bite down hard, preventing deeper introduction of food. Therefore, in this instance, small pieces of food should be used.

Once the food item is inside the mouth cavity, close the reptile's mouth gently with your thumb and forefinger, and hold it shut while you place the reptile back into its cage or upon a table top or floor. If the meal is immediately disgorged, repeat the operation, but this time keep the reptile's mouth closed over the intended meal when you put the specimen down. Another pair of hands may be needed if a large specimen is being handled to stop the animal from writhing about or otherwise moving. If it makes the procedure any more comfortable for you, release your finger from the specimen's chin but maintain firm but gentle pressure downwards on the snout in order to keep the mouth closed over the meal. After a while, when the reptile is motionless, very slowly release your hold on its body. If, when free of your restraint, it doesn't resume writhing, slowly relinquish your hold upon its head and back away. If undisturbed, it should then finish swallowing the meal of its own accord.

Force-feeding

If you are unsuccessful with the assist-feeding technique described above, you may need to consider forcing the food item a bit further down the reptile's oesophagus. Force-feeding is really best carried out on snakes and goannas, and the following description of the technique is intended for these only. You first need to open the recipient's mouth, and this is best done by gently nudging the mouth with the edge (not tip) of a smooth, lubricated metal rod or similar object (an unsharpened pencil will suffice for larger specimens) between the reptile's 'lips'. Once the snake or lizard's mouth is opened, work the rod towards the angle of the jaws. The mouth will then be completely opened, and you will be able to see the various structures within, and thus avoid mistakenly forcing the food item down the wrong passageway. A major concern is the possibility of the food item getting caught up in the opening of the trachea (larynx), or if a snake is being force-fed a rodent, ensuring that the lower jaw of the food item doesn't open, as its prominent incisors can damage the snake's mouth and oesophagus.

The food item, which should be lightly coated with an edible lubricant such as vegetable oil, egg yolk or margarine, can now be placed in the reptile's mouth, introducing it snout first over the rod and into the oesophagus. Once this is done, slowly slide the rod out sideways and use the end of it or a pair of long forceps to gently push the prey item down the oesophagus, and towards the stomach. If it doesn't slide easily, don't force it. Check to see if the food is being caught up at the angles of the jaws or elsewhere. If the rod slips past the food item, withdraw the rod and reposition it to resume pushing the food towards the stomach - at least until it is out of sight.

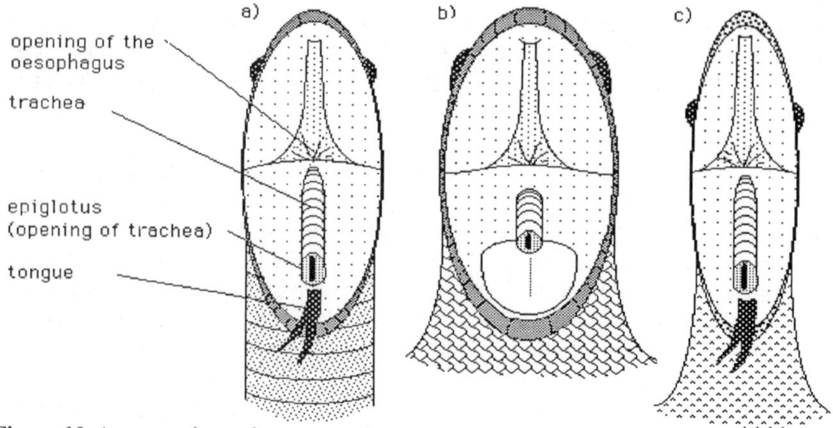

Figure 13 A comparison of structures in the mouth of a snake (*a*), a typical lizard [all but goannas] (*b*), and a goanna (*c*).

If a snake is to be force-fed a whole animal such as a rodent, choose a relatively small one. It is better to feed two or more small food items than a single one that involves excessive stress and strain upon the reptile. Whatever you use to push the food down the oesophagus with, make sure it is blunt and without sharp edges. After withdrawing the pushing device, gently manipulate the food item towards the stomach by running your thumb along the snake's ventral surface with your forefinger on the dorsum. This should be done to stop the snake regurgitating the food. When you have massaged the food item to a position of about a third of the way down the snake's body, put the reptile down and very slowly release the grip of your thumb and forefinger and step back. If it doesn't immediately attempt to regurgitate, wait until it begins moving forwards. At this time it should automatically continue the swallowing process. You can now return the force-fed reptile to its cage.

Tube-feeding

An alternative to force-feeding whole food items to a starving lizard or snake is to force a prepared food solution into its stomach via a long flexible tube attached to a large syringe or equivalent device. You may need to ask your veterinarian for help in obtaining a suitable device. As a rule of thumb, use a tube with a diameter of about a fourth of that of the snake, or for lizards, perhaps a bit smaller. This technique is also a useful way of administering drugs mixed in the food solution when a specimen is refusing to feed. It is particularly useful for involuntarily feeding sick snakes, and if done properly can be less stressful than the forced swallowing of whole food items. The 'stomach-tube' should be soft at the tip and free of sharp edges. Transparent aquarium air hoses are available in various diameters and are most suitable.

Tube-feeding begins with opening the reptile's mouth with a rod as described above. The stomach-tube should be lubricated with vegetable oil, raw egg, margarine or some other edible oil. Great care must be taken to ensure that the tube is inserted into the oesophagus, and not the trachea. NEVER begin tube-feeding without making 100% sure that the tube is correctly positioned. With snakes, a rule of thumb is to introduce the tube no more than one third the body length, while in lizards, you may only be able to advance the tube a short distance. Do not force the tube if it will not progress easily. You may need to withdraw it in order to add more lubricant. As in the case with snakes that are force-fed whole food items, you must make sure that the introduced food substance is not immediately regurgitated. Gently squeeze the body just anterior to the bulge created by the food as you withdraw the tube. Retain this grip around the body while putting the snake down. As soon as the snake begins moving forward, you can relinquish your grip and the food will move the rest of the way towards the stomach.

What to Tube-feed

If you are tube-feeding a normally carnivorous lizard or a snake, you can prepare a suitable mixture for tube-feeding. For the purist, appropriate whole prey items can be ground up in a high speed blender while frozen, and then thawed. Blending whole prey animals that are not frozen results in a gruesome runny mess. More conventional formulae include finely minced ground beef - appropriately supplemented with multiple vitamins; or a mixture of mashed, hard boiled eggs with enough water to provide adequate fluidity. Whatever formula you use, it should be as thick a consistency as can be forced through the tube, since it may otherwise ooze back out of the recipient's mouth.

Vitamin Supplementation

It has been demonstrated that tortoises and at least some lizards are subject to vitamin and mineral deficiencies when not exposed to sunlight or artificially produced ultraviolet light. These species normally utilize the ultraviolet rays to produce Vitamin D in their skin. Vitamin D plays a vital role in the uptake and utilization of calcium by the body. A deficiency in Vitamin D or calcium causes rickets and skeletal deformities in all reptiles, often seen as a softening and deformity of the shells of tortoises. Typical symptoms for lizards deficient in Vitamin D include deformed jaws, crooked tails, easily fractured bones and the production of malformed eggs or young. Snakes and many lizard species apparently get adequate amounts of Vitamin D through their usual diets. However, other reptiles normally require that Vitamin D be produced in the skin. Fortunately, it appears that in the absence of sufficient ultraviolet radiation, the oral ingestion of supplemental Vitamin D (especially Vitamin D3) and calcium can make up the deficit in most species.

There are not any hard and fast guidelines for supplemental vitamin and mineral doses, and it is difficult to make any precise recommendations. The danger of over-supplementation exists, which can result in as serious a health problem as a vitamin deficiency. The most common practice currently used by lizard breeders is to 'dust' a Vitamin D and phosphorous enriched calcium powder (bone meal) onto the insects - or whatever food item is to be fed to the lizards. The best way of 'dusting' insects is to place a small quantity of the powder into a transparent plastic bag with the insects. Then blow up the bag and shake until the powder coats the insects.

A pinch or so of Vitamin D enriched calcium powder can also be mixed with a tortoise's food, but ensuring the tortoise has ingested the powder before it is washed off in the water is a problem. You may need to use some ingenuity to get the powder inside bite size morsels that your tortoises will swallow whole. It is possible to mix the vitamins with minced meat which is then frozen. When the tortoise is to be fed, small pieces can be broken/cut off, thawed slightly and dropped into the water.

This supplementation should not be given at every feeding for lizards or tortoises, but perhaps only once a fortnight (certainly no more than once a week). It should be given in association with a healthy, well balanced diet to ensure the introduction of other essential minerals (particularly phosphorus). The inter-relationships between vitamin D, calcium and phosphorus are quite complex and beyond the scope of this book.

Some reptile keepers place drops of multivitamins such as Avedrops (sold at pet shops for birds) in the water for their lizards (after withholding water for a day or so previously to ensure thirst), or inject a small amount into food items intended for their tortoise. As long as this supplementation isn't overdone, it may be quite useful in contributing to the health of your reptiles. Again, there are as yet no known 'correct' dosage rates, but a few drops in the water bowl (offered once a week or fortnight), or one or two drops mixed in an adult tortoise's food item (a drop or less for juveniles) on a once a week basis should be safe.

Skin Sloughing

Snakes and lizards all periodically slough (shed) the outer layer, (epidermis) of their skin following the production of a new one beneath it. Younger individuals slough more often, and the more rapidly the reptile grows, the more frequently it will slough. During the early stages of the skin sloughing process you may notice your reptile's colour beginning to darken and lose its lustre. This condition usually lasts a few days to about a week and is brought about by the production of the new epidermis beneath the old. With the completion of this stage an oily, substance is secreted between the new and old layers of skin, freeing and separating the old skin. During this time the skin takes on a milky appearance, particularly on the underside, and in the case of snakes, over the eyes. During this 'opaque' condition, reptiles should not be handled or otherwise disturbed because the skin is particularly vulnerable to damage. In a few days this milky appearance will disappear and the reptile will be ready to slough. You may notice it moving about the cage, rubbing itself against whatever objects are available to assist in peeling the old skin off, either in large pieces (most lizards) or in the case of snakes, generally the entire skin in one piece. Snakes begin the peeling process at the snout and finishing at the tail (turning the cast skin inside out in the way a stocking is removed from a foot). Some lizards eat their cast skins. Tortoises also occasionally slough sections of the outer (epidermal) layer of their shells.

Sloughing Problems

Sometimes captive snakes and lizards develop problems during the sloughing process, and may require some help. If a week to ten days has passed following the conclusion of the 'opaque' stage, and the reptile has failed to slough its skin (or any large portions thereof), you should intervene. Place the reptile in a sealable, water-tight container such as a clean plastic rubbish bin with a fastenable lid. Add luke-warm water to a level that would just about submerge the reptile's body. It will be able to hold its head above water to breathe and you needn't worry about it drowning. You may find after an hour or so that the problem skin is beginning to come loose on its own accord, and if you return the snake or lizard to its cage it will manage to complete the job itself. If on the other hand, after several hours of soaking the skin still shows no sign of loosening, try gently rubbing it loose (not venomous species - the risk of being bitten outweighs the advantage of assisting in the sloughing). Don't force it if it doesn't loosen fairly readily. In the case of non-venomous snakes, begin with the skin on the snake's snout and chin, rubbing it in the direction of its tail to encourage the start of the peeling process. If you are very careful, and the skin easily peels, you can work the skin backwards to say, a few centimetres past the head, which should prompt the snake to finish the procedure once it is returned to its cage.

A word of warning should be made regarding the forced removal of unshed skin from a snake's eye. The transparent skin that protects the eye in its ever open condition is called the 'spectacle'. If the old spectacle doesn't remove easily you could damage the eye permanently by forcing the new one off with it. In any event, the old spectacle will most probably be removed with the next sloughing cycle, and presents no immediate health problem to the snake. Spectacles that are left behind may be an indication of an infestation of snake mites, which bore through the edges of the spectacle, isolating it from the rest of the snake's outer skin (more on snake mites in diseases section).

Diseases

In general, treatment for any but the most routine medical problems that your reptiles experience should be carried out by an experienced veterinarian. If you don't know of one with experience in treating reptiles, contact a nearby fauna park or your herpetological association for the names of known suitable veterinarians in your area.

Stress and Captive Reptile Diseases

For a captive reptile, stress can be defined as the cumulative response to abnormal influences in its environment conveyed via its senses. Excessive stress can weaken a reptile's defences, leaving it susceptible to disease. Excessive handling, disruption of the day/night cycle, and too cold or too warm conditions can all contribute to the degree of stress experienced. The psychological stress of being captured and the frustration of not being able to escape are likely to occur in newly captured specimens, but this can be minimized if suitable housing conditions are provided, including secure hiding places and lots of quiet. In treating a sick reptile, a balance must be struck between handling for treatment and leaving the animal undisturbed. All handling of sick reptiles is stressful.

The Common Forms of Bacterial Diseases in Captive Reptiles

Infectious species of bacteria are conveniently divided into two major groups: 'gram positive' bacteria and 'gram negative' bacteria. For our purposes, the significance of this distinction lies in the fact that gram positive bacteria are generally associated with the diseases of mammals and birds, while infectious reptile diseases are almost always caused by gram negative species, (usually *Pseudomonas* or *Aeromonas* species). Veterinarians with experience in treating reptiles will be aware of the types of antibiotics available and the dosage rates for treating any severe bacterial infections that your reptiles might incur, but it is important that you understand that the usual cause of these infections is inadequate husbandry (usually insufficient heat and/or unhygienic conditions). The most frequently occurring bacterial diseases amongst captive reptiles are: respiratory infections (usually referred to as 'pneumonia'); canker (also called 'mouth rot'); scale rot or blister disease in snakes and lizards; osteomyelitis (also called 'shell rot' in tortoises); local abcessation in lizards; and panophthalmitis (a progressive infection involving the eye in snakes). Whilst the causes and symptoms of these diseases overlap to some degree, they are considered here individually.

Respiratory Infections

Respiratory infections will generally afflict lizards or snakes that are kept in excessively cold or damp conditions and may also occur in tortoises kept too cold or incorrectly. The symptoms can include audible breathing difficulty - which in the early stages sounds like 'wet hissing', progressing later to 'sneezing', wheezing and a ballooning of the throat area during exhalation. Also symptomatic of respiratory infection is a lack of appetite and the presence of copious amounts of mucus in the mouth or nose. A common posture adopted by lizards or snakes with respiratory infections is an elevated head, often supported in a corner of the cage or on a water dish, presumably to aid in breathing. If recognized in the early stages, adding warmth and dryness may be sufficient to bring about a speedy recovery. Otherwise, diagnosis by a veterinarian is advised, with treatment generally calling for a course of antibiotics.

Canker or 'Mouth Rot'

Canker, or 'mouth rot' as it is sometimes called, occurs all too frequently amongst snakes and lizards kept in less than optimal conditions, and it can be recognized in its early stages by a swelling of the tissues in the mouth. This inflammation prevents the mouth from closing properly. Inspection of the inside of the affected reptile's mouth, particularly along the gums, will show the presence of grey-white or yellow-white 'cheesy' matter clinging to or replacing the mucus membranes in the mouth. This is not to be confused with the thick, yellow-white mucus present in cases of severe respiratory infection, and is often preceded by small reddish spots (haemorrhages) on the gums. The first step in treating this disease is to rectify the caging circumstances that were responsible for its onset. Treatment of the snake will need to be supervised by a veterinarian and consists of: carefully scraping away the cheesy matter from the mouth daily, followed by a rinse of the mouth with water or 3% hydrogen peroxide then application of an antibiotic ointment to the infected parts and finally a course of antibiotic injections. Vitamin C supplementation is known to promote tissue healing, and may therefore be a useful adjunct in the treatment of canker. There is strong opinion amongst hobbyists (the author included) that the continuous scraping away of the cheesy matter is of limited value in mild cases.

'Scale Rot and Blister Disease'

Both 'scale rot' and 'blister disease' are phases of the same disease. This highly contagious reptile disease occurs in lizards and snakes, but is more common in snakes. It is promoted by a combination of poor cage sanitation and excessively cold, damp or humid conditions and often results from the practice of placing reptile cages on a cold floor with insufficient heating supplied only from above. It can also occur as a result of having live plants in the vivarium without proper consideration for dryness and sanitation. The disease often first appears as 'scale rot', when patches of ulcerated skin appear, usually on the reptile's underside; or as 'blister disease', with a series of blister-like lesions appearing between and beneath the scales. In the earliest stages of scale rot the ulcerations are reddish-brown areas which can be quite large. As the ulcerated patches deteriorate, they become a series of open sores that can extend along the entire length of the reptile's underside. In the 'blister disease' phase of the disease (which may or may not follow a 'scale rot' phase) the blisters can quickly spread to cover much of the animal's body, and the blisters are filled with fluid that can be either reddish, clear, or milky. Blisters will often spread rapidly in the 'opaque' phase of shedding.

As is the case in nearly all reptile diseases, it is imperative that treatment begins immediately following diagnosis. If detected in the very early stages, a simple matter of remedying the improper caging conditions and disinfecting affected skin may be sufficient treatment. Otherwise, veterinary treatment with antibiotics will be necessary. Areas of skin discolouration may persist following healing, but these will become less pronounced with each successive sloughing of the skin. If left untreated the scale rot phase of the disease will progress until the ulcerations eventually perforate the body wall causing death. Death from the blister disease phase can occur more suddenly and without warning.

Osteomyelitis of the Shell or 'Shell Rot'

Tortoises are susceptible to a condition commonly known as 'shell rot', a form of osteomyelitis of the shell. It is very often mistakenly considered a fungal infection, but in reality is a bacterial disease. The plastron (underside of shell) is usually most affected and characteristically the ulcers form large yellowish patches. If untreated these ulcerations will slowly progress and become edged with red.

The disease usually begins with a simple abrasion to the shell (which most often occurs by continuous scraping of the plastron on excessively rough edges of concrete - especially in ponds where the concrete is uncured) which then becomes infected. This early scratching may cause superficial bleeding and therefore be red. The infected area of the shell becomes soft and is easily scraped free from the underlying healthy shell. Over time the lesions will expand slowly and erode deeper into the shell; eventually perforation of the body wall will lead death.

Treatment will need to be supervised by a veterinarian and consists of: a) keeping the tortoise in a warm, dry place, except for an hour a day when it is placed in clean water in order to drink or feed; b) a course of antibiotic injections; c) removing the affected 'scabs' daily - a soft bristle tooth brush can be used for this purpose; and d) coating the exposed area with an antibiotic ointment.

Treatment of this disease can be very lengthy since the shell is slow to heal. Don't be discouraged if it takes more than a month to resolve the problem. During the period of treatment any rough surfaces in the tortoise's pond that may have contributed to the onset of the disease should be attended to.

Local Abscesses in Lizards

This condition is common in bluetongues, especially following apparently superficial bite wounds from cage mates, or in those animals kept in dirty caging conditions. Localized abscesses commonly occur on the feet (where they may involve toe loss or even foot loss), on the head, or on the back. Poor nutrition often promotes the spread of the disease. Treatment consists of removing the hard, cheesy matter, and the surrounding fibrous capsule, applying antibiotic ointment, and maintaining conditions of cleanliness, warmth and dryness.

Panophthalmitis

Panophthalmitis occurs primarily in snakes, and is an infection of the tissues underneath the spectacle, which is that part of the snake's skin that covers the eye. It is probable that bad cage hygiene and/or cold or damp conditions lead to the onset of this disease though it is more likely to arise following a bout of canker which has caused the blockage of the lacrymal duct (the lacrymal duct leads from the roof of the mouth to the eye above); in either case, causing infection within the duct. The first sign of panophthalmitis can either be an enlargement of one of the snake's eyes, or the clouding over of the eye with or without swelling. The appearance is similar to the 'opaque' condition of the eye which occurs during skin sloughing - except of course that the other eye remains clear. As the disease progresses, the eye may become grotesquely swollen. Sometimes the eye will be 'blood shot' with small blood vessels criss-crossing the discoloured portion of the afflicted eye. In reality, most of the swelling and infection takes place in the space between the eye itself and the spectacle.

Panophthalmitis must be treated by a veterinarian, who if possible will remove any visible blockage of the affected lacrymal ducts where they terminate at the roof of the mouth. The condition is often improved dramatically by a course of antibiotic injections. In more advanced stages the spectacle may need to be surgically removed. Unfortunately, since this leaves the eye unprotected, it too must be removed.

Internal Parasites

Reptiles have co-evolved harmoniously with their parasites. In the wild, this balance can sometimes be shifted, leading to the death of the host, but generally the host and its parasites live in a state of balance. This balance is often upset in captivity as a result of undue stress which can act to break down the reptile's resistance.

Since most of the potentially dangerous internal parasites depend upon intermediate hosts prior to infesting your reptiles, the feeding of 'laboratory' reared prey, or artificial diets, can minimize the introduction of at least some of them to your animals.

Reptiles harbour numerous intestinal parasites, some of which are virtually harmless while others can cause severe disease. There are various drugs that can be used for treatment of the more dangerous parasites, but each drug is generally only useful in treating specific types of internal parasites, and most are potentially toxic to reptiles. It is economically and medically unwise to treat reptiles for diseases that have not been positively identified. Instead of contemplating 'shot-gun' remedies for reptiles merely suspected of suffering from internal parasites, it would be prudent to provide a faecal sample for examination by your veterinarian who will then be able to recommend specific treatment should the tests prove positive for protozoa or worms. A follow-up faecal examination three to four weeks after treatment will show if the treatment has been effective.

Most drugs recommended for treating protozoan infections are given orally. This is best accomplished by inserting the tablet or portion thereof into the centre of an intended food item. If the preparation is in powder form, mix to a slurry with a small amount of water and inject this into the food item. Fluid preparations can likewise be administered. If the reptile refuses to eat or regurgitates the treated food item, the drug may be mixed with water and tube-fed to the reptile.

Protozoan Infections - General

There are a multitude of protozoa that infest reptiles both in the wild and in captivity. Most are unlikely to cause severe health problems in your captive specimens. The problem ones cause gastrointestinal diseases. They can strike regardless of cage hygiene or temperature, and should be regarded as highly contagious to the rest of your reptile collection. The most commonly encountered of the potentially dangerous ones are discussed here briefly.

Amoebae

Early symptoms of infestations by amoebae [amoebic dysentery] are subtle - in fact death can be the first indication of an infection. Signs to look for include: general lethargy, regurgitation of meals or an unwillingness to feed, a seemingly endless thirst for water, and in some cases, watery-green or bloody diarrhoea or constipation. In snakes a constipational blockage caused by amoebic infections can be felt as a large, very hard lump - usually between 3/4 and 9/10 of the way down the digestive tract. In later stages of amoebic dysentery snakes characteristically coil in a very loose manner, and don't move much.

The presence of the amoebae can be confirmed by the inspection of a faecal sample by a veterinarian. Treatment with Flagyl (orally) is straightforward and highly successful if commenced before substantial damage to the digestive tract has occurred. As an adjunct to drug treatment, warming the affected reptile somewhat for a period of a week or so can be valuable. A follow-up examination of faecal samples following treatment is recommended.

The most common way that this disease enters a collection is through the practice of feeding infected but often apparently healthy food items (such as frogs or skinks) to snakes. There is a real danger of spreading the disease to cage-mates via direct or indirect contact with faeces or regurgitated food. The amoeba - Entamoeba invadens - generally does not naturally occur in all snakes and lizards in the 'benign' state, as it does in some of their food items. The disease is highly contagious in a reptile collection, and good cage hygiene, as well as the complete isolation of infected individuals, is very important.

Flagellates

Various types of flagellates can cause diseases such as necrotic enteritis in captive snakes and lizards. They are so named because they propel themselves through their fluid environments by waving whip-like appendages called flagella. Outbreaks of intestinal infections of flagellates are often prompted by inadequate caging conditions or undue behavioural stress such as that caused by the overcrowding of territorial species. Some flagellates (eg trichomonads) are so contagious that if introduced into a collection, they can spread regardless of cage conditions. The

left The long-necked tortoise *Chelodina longicolis* soon learns to accept food from fingers.

below An estimated five million reptiles and frogs are killed each year on Australian roads, paling to insignificance the small number collected by hobbyists.

bottom A western brown snake *Pseudechis nuchalis*.

left The longest of Australian goannas, the perentie *Varanus giganteus* is an active desert species requiring large quarters and warm dry conditions if it is to b kept in captivity.

right The function of th unusual 'knob' belongin to knob-tailed geckos (genus *Nephurus*) is not known.

left Dragons such as the colourful northern specie *Diporiphora magna* are sun lovers, often seen basking on rocks, fallen trees, and other vantage points.

right Able to swim through sandy desert soils, the broad-banded sandswimmer (*Eremiascincus richardsoni)* emerges at night to hunt for invertebrates and smaller lizards.

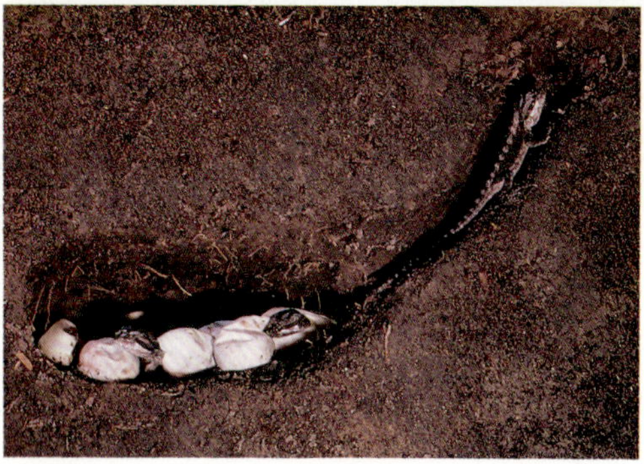

ɔove An undescribed species of short-ːcked tortoise from northern NSW.

ɔove right, right A yellow-faced whip ıake *Demansia psammophis* sloughing s skin.

·ft Classified as an endangered species, ıe broad-headed snake *Hoplocephalus ungaroides* may not be collected from ıe wild.

ight The Tryon's ecko *Oedura tryoni* a large (to 16cm) nd colourful species ving mainly mongst rock utcrops. Adjusting eadily to captivity, it oon becomes a erocious feeder.

ır left Bearded dragons *Pogona barbata* hatching from their eggs before digging ıeir way free.

·ft The smallest of python species, the dwarf Children's python *Bothrochilus erthensis* attains an average length of only 30cm.

left Most reptiles retreat for extended periods within burrows, hollow branches, or beneath suitable objects. Here a specimen of *Drysdalia rhodogaster* hides beneath a fallen piece of bark.

below Some reptiles rely upon camoflauge to avoid detection by predators. Here, a spiny-tailed gecko *Diplodactylus ciliaris* melts from view amongst the mottled colours of a tree trunk.

bove The common tree snake *endrelaphis punctulatus* exists in a ariety of colour phases including his rare blue form.

ight The spectacular jungle carpet ython *Morelia sp.* is confined to the icinity of the Atherton tablelands in opical Queensland.

right Although commonly kept and bred in overseas collections, the green tree python *Chondropython viridis* is uncommon in Australian collections.

below A diamond python *Morelia spilota* with her clutch of eggs. Most hobbyists choose to artificially incubate python eggs rather than leave them with the brooding female.

below The pink-tongued skink *Tiliqua gerrardii* feeds mainly upon snails and slugs in the wild, but will accept canned pet food in captivity. If snails are to be fed, they must be collected away from areas where snail-bait is used.

symptoms of a flagellate infestation are similar to those described for amoebic dysentry (see above). As is the case with amoebic infected individuals, you should isolate the reptile, and be careful not to contaminate other cages, particularly by using tools to clean uninfected cages that have been used to clean infected ones.

Early symptoms of a flagellate infection include loss of appetite, and in lizards, the presence of remnants of faeces around the cloaca or under the tail. In some cases, particularly in Trichomonas infections, regurgitation follows feeding. Your veterinarian will be able to identify most infections upon microscopic inspection of faeces. Treatment generally involves the oral administration of Emtryl. If the reptile can be trusted to drink regularly, this drug can simply be added to its drinking water.

Other Protozoa

Other infective protozoa are less frequently encountered. One of the most dangerous is Cryptosporidium, which in recent years has been the subject of study by the reptile keepers and veterinary staff at Taronga Park Zoo. The diagnosis of cryptosporidiosis requires specialized techniques, and although it can be controlled with drug therapy, there is no known 'cure'. The symptoms of cryptosporidiosis include regurgitation of meals (which may be covered with excessive slime), defaecation of undigested meals, substantial weight loss and a visible mid-body swelling. Symptoms can persist for months or even years before death occurs.

Round Worms

Round worms (nematodes) are unsegmented and cylindrical in cross section. Some types of round worms inhabit the digestive tracts of reptiles, and in captive circumstances can reach dangerous numbers. Because some attach themselves to the gut wall, local ulcerations that become infected can occur secondarily. You may be able to see some varieties (ascarids) in the faeces of your reptiles in moderate numbers. Worm eggs will be readily detected by your veterinarian upon faecal examination. Ascarid nematodes are readily eliminated with Piperazine given orally at a dosage rate of 110 mg/kg. Reptiles infested with 'hookworms' should be treated with Mebendozole at a dosage rate of 50-100mg/kg.

Tapeworms and 'Skin Worms'

Tapeworms are parasitic flat, ribbon-like worms. Adults live in the gut of the reptile and their long ribbon like body consists of a chain of segments. It is in this adult stage that tapeworms are generally detected in the digestive tracts of reptiles. Like round worms, they can become exceedingly numerous in captive reptiles, and then cause ulcerations of the gut wall. They also sap the reptile of its energy, and in large enough numbers, can cause malnutrition or even intestinal blockages. Faecal examinations by your veterinarian will determine the presence of tapeworm eggs or segments. The popular treatment for reptiles with tapeworms is Droncit at 2-7 mg/kg (smaller animals get proportionately higher doses), given orally.

The larvae of some tapeworms are commonly found in cysts beneath the skin in certain snakes - particularly frog-eating species (eg. green tree snakes). The cysts are harmless, but for aesthetic reasons they may be removed via a small skin incision. Following incision the worm is squeezed partially out before being grasped and removed with a pair of forceps.

Flukes

Flukes are parasitic flatworms also known as trematodes. Adults may live in the liver, gut, lung, or blood-vessels of the reptile, where they may cause serious disease. Swamp-inhabiting snakes such as tiger snakes, copperheads and swamp snakes are often infested with 'lung flukes' upon their capture. Treatment is difficult since the killing of the flukes can result in more serious problems when they lodge in the respiratory tracts.

External Parasites

The most important external parasites that afflict captive reptiles are mites and ticks, closely related organisms of the same class as spiders (Arachnida).

Mites

The snake mite *Ophionyssus natricis* is regarded by some reptile keepers as public enemy number one. It mainly affects snakes but also some lizards, particularly those with overlapping rather than nodular or conical scales.

Mites like warmth and darkness. By day they hide away either in the tiniest imperfection in the cage floor or under the scales of the host, while at night they emerge to find a suitable position on the reptile's skin to suck blood. Under favourable conditions they breed when a few weeks old. A mite full of blood can fast for 14 days or more. The nymphs hatch out from the eggs in 10-35 days, depending upon the temperature. The larger mites and even the nymphs creep into other terraria and in no time at all can infest an entire collection. They are particularly dangerous to small or juvenile specimens, which can be tormented and bled incessantly.

There is some speculation that mites may also act as a disease vector, transporting infective agents from one reptile to another, but this has yet to be proven. Once these tenacious parasites are introduced into a collection, they are difficult to completely eradicate, tending to reappear as outbreaks from time to time. They can be somewhat controlled though, by keeping cage conditions exceptionally clean, and by replacing any substrate materials regularly. They are much easier to control where paper cage substrate is used in preference to sand, leaves or other materials that provide plenty of hiding places for the mites and are less likely to be changed often.

You cannot expect to easily see mites with the naked eye because they are so small, though when gorged with blood they may appear as tiny black or reddish-brown specks behind the scales. If kept on a paper substrate, blood gorged mites will be easily seen after dropping from their host. If kept otherwise, the best way to detect them is by looking for the tell-tale signs of their presence such as the typical inflammation of the area around the eye. An unnatural gap will be present between the eye and the surrounding skin in badly infested specimens. This does not always apply to large pythons though, since a similar gap may exist even when no mites are present. Inflammation beneath body scales (in both snakes and lizards) caused by mites can push the scales up into a somewhat raised condition.

Another sign of infestation in either snakes or lizards is their tendency to soak in their water bowls for lengthy periods of time. In such cases, you may be able to actually see tiny blood gorged mites, presumably drowned, at the bottom of the bowl. In the absence of water to soak in, the reptile may be constantly on the move, rubbing against the sides and contents of the cage. A secondary problem of respiratory infection, scale rot, or canker is likely to eventually arise in reptiles (particularly pythons) that spend excessive time in cold water bowls due to mite irritation.

Sometimes the only sign of infestation is the presence of mite droppings - visible as tiny white specks on the scales. If you handle a mite infested reptile, your hands and arms may begin itching, following the transfer of some of the mites to you. If you subsequently handle other reptiles or their cages, you will spread the mites to these. Fortunately a hot shower and (and clothes washing) will remove mites from your body should you detect them.

It is believed that *Ophionyssus natricis* is not an Australian native species, but was brought into the country with exotic reptile imports. Mite infestations may enter a collection whenever an infected specimen from another collection is brought into it. There is some debate as to whether any Australian snakes in the wild have mite infestations. As a precautionary measure, all new acquisitions obtained from other keepers should be carefully inspected for signs of mites and kept apart from the rest of your collection until given the 'all clear'. Some hobbyists treat all of their new arrivals for mites irrespective of whether or not they are able to detect any signs of infestation - just to be sure.

There are several popular treatments for mite infested reptiles - each with its associated advantages and disadvantages. All can be considered highly dangerous to the health of the reptile if improperly administered.

At the time of writing, the most promising of these treatments consists of 'fogging' the affected cages (with reptiles still within) and the room within which the cages are maintained with an aerosol spray can of 20g/kg d-phenothrin 20:80. This is the product used to spray the interiors of arriving international flights. Unlike the chemical dichlorvos (contained in 'Shelltox Pest Strips', see below), d-phenothrin is rapidly biodegradable and is not stored in the body tissues of those exposed to it. This chemical is far less dangerous to human beings and reptiles alike than dichlorvos (which has proven to be cancer-causing) and to date has not been held accountable for any reptile fatalities. It is recommended that any food, water, or empty water bowls be removed from areas being sprayed. There is as yet no firm recommended regimen of spaying, but as with other mite treatments, it should be used at least twice (perhaps the second spraying a few days to a week after the first) to ensure effectiveness. It is recommended that spraying be done at night as this is when mites are most active and are less likely to escape contact with the chemical. This product is not widely marketed, and you should contact your herpetological association for details on how to obtain it.

The most popular mite treatment is the 'Shelltox Pest Strip'. The small blue form should be used for cages, the larger yellow form for treating an entire room of cages or larger enclosures. This product is widely available at gardening supply shops and supermarkets. To prevent the reptile contacting the Pest Strip, place it in a sealed paper bag with small holes throughout (or some other perforated container) and put this in the cage overnight only. If you are treating the entire room you may need to use more than one of the large yellow strips at once, or leave the strip out for longer. All drinking water should be removed during treatment, since submerged reptiles will not be exposed to the pesticide, and cages must have reasonable ventilation. The treatment should be repeated in three or four days time in order to kill any mites that survived the first treatment. The active chemical in Pest Strips (dichlorvos) is very toxic to reptiles, and has often been known to kill reptiles treated with it. For some unknown reason, this is particularly true with death adders, young snakes and small lizards. If any of these are to be treated, it is imperative that adequate air circulation exists, and it is probably a good idea to treat these for a much shorter duration - say three hours at a time.

Another means of dealing with mites is to 'dust' suspected carriers with 'Dri-Die 67' powder - an abrasive chemical that kills the mites by scratching their chitinous shells causing dehydration. This method is considered by some reptile keepers to be the most effective means of eliminating mites and keeping them out of a collection, but others suspect that it causes excessive drying out of the reptiles' skin. Dri-Die is used by aviculturists to control bird lice and is more difficult to obtain than Shelltox Strips. If you use Dri-Die avoid inhaling it, and try to minimize the amount inhaled by the reptiles being treated.

Yet another popular treatment for mite infested reptiles is to spray them and their cages with a solution of Neguvon, a veterinary product, from an indoor plant mister. This product is very toxic, and it should be used with great caution. Many fatalities of reptiles treated in this manner are known. Nevertheless, the reptile keeping staff of Taronga Park Zoo have found that when used carefully, Neguvon can in fact be a highly effective and relatively safe remedy for mite infestations. The procedure which they have adopted is to mix the chemical (wearing rubber gloves, and exercising care not to inhale the powder) at a rate of two teaspoons per eight litres of water. Snakes and lizards (excluding lizard species that take up water through their skin) can be dipped into the solution up to their head. Eye orbits and other folds on the head can be wiped with a cotton bud dipped in the solution. While the reptile is out of its cage, the cage can be thoroughly cleaned, all cage props disinfected, and the insides of the cage sprayed with the Neguvon solution. After an hour or so, both cage interior and reptile(s) should be thoroughly rinsed, and when the cage is dry, the reptile(s) returned.

Ticks

Wild caught snakes and lizards very often have many ticks embedded beneath their scales. These can be daubed with alcohol, then easily removed. If not removed, ticks will eventually drop from their host, but may then either lay eggs, or pass through a metamorphosis only to reattach to the reptile.

Intestinal Obstructions

Intestinal obstructions and constipation are conditions occasionally occurring in captive reptiles, though one must remember that many species of lizard and snake (pythons in particular) can go for very long periods of time without defaecating, regardless of whether feeding is continuing. Blockages of the digestive tract most often occur between the stomach and beginning of the colon, and can be caused by dehydration or improper diet. This often occurs when heavily furred or feathered animals are fed constantly to large snakes, or when mealworms or other highly chitinous insects are supplied as the sole food/diet to lizards. Other causes of obstruction include protozoan infections and excessive parasite burdens. Intestinal obstructions and constipation occur less commonly in tortoises.

Unfortunately, symptoms of intestinal blockages, other than a failure to defaecate, may not be apparent even though damage to the intestines or colon has occurred (secretions of white uric acid deposits and bile material will continue to be eliminated from time to time). Symptoms that often appear include loss of appetite, lethargy, or the presence of gas-filled loops in the intestines causing a bulge in the body. Constipated snakes often adopt an unusually straight position when resting, tending not to coil in the typical fashion.

If you suspect constipation or an intestinal obstruction in one of your reptiles, take the specimen to your veterinarian for inspection. If it is still feeding, he may advise you to feed it small food items that have been injected with an edible oil, such as a vegetable oil; or he may tube-feed it oil directly. Enemas may help and can be used in association with gentle manipulation of the blockage. This step should definitely not be attempted without the veterinarian's help, since damage to the affected area can easily occur, particularly if it is secondarily infected. Antibiotics are likely to be needed in follow-up treatment.

Additional Health Disorders in Captive Tortoises

In addition to 'shell rot', which has already been discussed, there are several other medical disorders commonly encountered in captive tortoises.

Cracked Shells

It is common to find tortoises crossing country roads that pass near ponds, creeks or swamps during or after heavy rains. Unfortunately, quite often these are hit by a passing vehicle and left with a fractured shell. These injured specimens can very often be saved if taken to your veterinarian for treatment. The wounds can be treated, and the shell fitted together - and either glued into place using Superglue or partially reconstructed with fibreglass. Antibiotic treatment is recommended to prevent infection.

Fungal Infections

The first indication of a tortoise being afflicted with fungal infections of the skin is the appearance of white to cream spots on the skin. These infections usually start around the claws or eyes, from where they spread rapidly if not treated. This condition often appears in newly captured tortoises placed in water of inappropriate temperature, pH or cleanliness. As mentioned earlier, it is essential to invest in a pH testing and adjustment kit if you are going to keep tortoises and they are inexpensive and readily available at pet shops and swimming pool supply shops.

The affected area of skin should be painted with an anti-fungal drug such as mercurochrome and the tortoise should be kept dry for at least an hour following treatment. This treatment should be given twice daily until the condition clears.

Swollen Eyes

This ailment is due to unclean conditions or low water temperature. The pond or aquarium should be thoroughly cleaned and the temperature of the water maintained at about 25°C. In severe cases an antibiotic cream or ointment should be applied following veterinary consultation.

Breeding

One of the major goals of keeping wild animal species in captivity should be to encourage or induce them to reproduce. Not only is captive breeding a measure of successful husbandry, but it lessens the demand for the collection of additional specimens from the wild. Only well-adjusted and healthy specimens will breed in captivity. Underweight or maladapted animals will have little interest or capacity for reproduction. Breeding can be regarded as an indication that the physical and psychological requirements of the species being kept are being adequately met and represents a gratifying achievement for the reptile hobbyist.

Breeding Tortoises

The sexes of tortoises are virtually indistinguishable as juveniles, but adults of most species can usually be sexed by an experienced keeper. Male short-neck tortoises (such as the Macquarie tortoise) have proportionately longer and stouter tails than do females. Male long-neck tortoises and their relatives (such as the broad-shelled tortoise) can *sometimes* be distinguished from their female counterparts by the possession of a slightly concave plastron (underside of the shell). This may assist in allowing them to remain in place above the female while mating.

You can only expect to breed tortoises when they are kept in outdoor facilities. A softened mound of soil maintained adjacent to your tortoise pond will provide a suitable egg laying site, though you will have to watch all soil areas of the enclosure for signs of disturbance during the period when egg laying is due if you intend to uncover and artificially incubate the eggs. The technique of incubating reptile eggs is discussed below. Most tortoise eggs, if kept at about 28°C, will hatch in approximately two to three months, though those of the broad-shelled tortoise may take longer. Eggs left to hatch outdoors are likely to take much longer, and if the egg laying site is not at least slightly moistened, the young may die because they will be unable to dig free of the hardened soil.

Hatchlings may be placed in a container with shallow water and a rock to climb out on. They should commence feeding about a week after hatching, and should then accept tiny morsels of chopped prawns or fish.

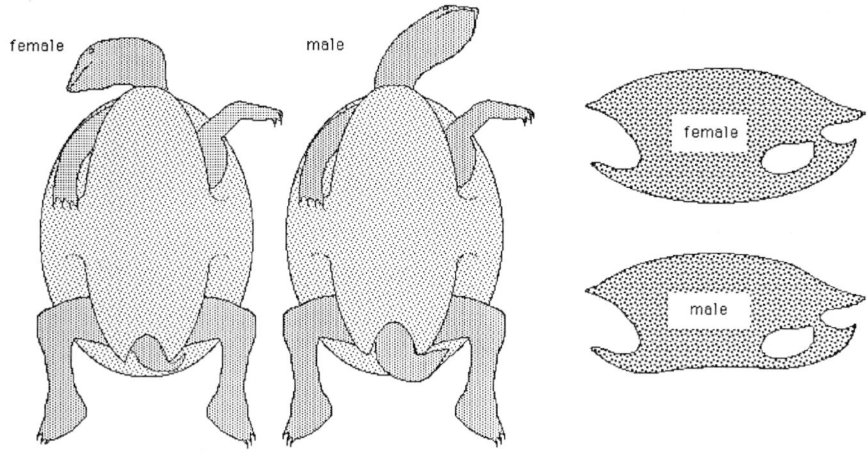

Figure 14 Adult short-necked tortoise species may be sexed on the basis of relative tail size *left*. A difference in shell shape can sometimes allow sexing of long-necked varieties *right*.

Identifying the Sex of Lizards and Snakes

Before you can expect any breeding success with your lizards or snakes, you should attempt to determine their sex. Sexing them can be rather difficult because of the absence of any obvious differences between the sexes of most species. Amongst the adult males of some species, a visible bulge is created along the base of the tail owing to the presence of the paired copulatory organs - called hemipenes (singular: hemipenis). In the species where this presence is most apparent, the female's tail tapers much more dramatically from the vent towards the tip than does the male's (see figure 15). In adult male geckoes, these protrusions lie just behind the vent and are quite obvious.

Fortunately, a technique exists to determine the presence or absence of hemipenes in snakes and large lizards. In males, a thin blunt-ended probe can be inserted GENTLY (after lubrication with vaseline jelly, KY jelly, or some other non-toxic material) into the vent, then into one of the hemipene pockets extending towards the tail tip.

The thickness of the probe should vary with the size of the specimen being sexed. For snakes greater than 60 cm long, the probe should be approximately the diameter of a wooden match, while for smaller specimens something proportionately smaller will be required. The probe must be either metal or smooth plastic and have a very blunt tip. Your veterinarian will probably be able to supply you with suitable instruments. Smaller probes can be constructed of various gauges of nylon monofilament fishing line, with the tip rounded by placing it just close enough to a lit match to melt away any roughness.

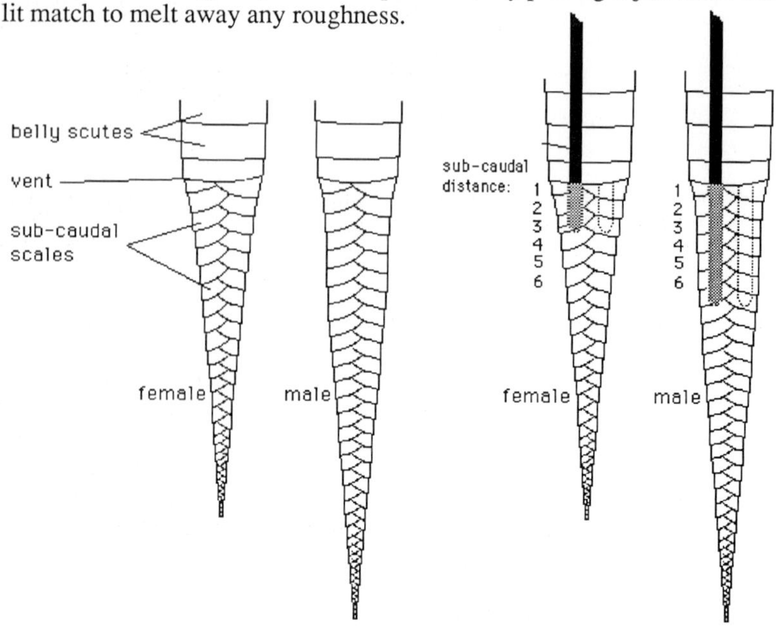

Figure 15 The sex of snakes and some lizards can be determined by an examination of the shape of the tail, or by using a blunt surgical probe (*right*) to establish the presence of the male's hemipenes. Tail shape variation is due to the bulge of the hemipenes. In some species, the tail is relatively longer in males.

Generally, in females, the probe can only be inserted to a distance equivalent to between one to four of the subcaudal scales (the scales on the bottom of the tail), while in males the probe should descend easily inside the hemipenis for a distance of at least six, usually more like eight to ten, subcaudal scales. To check the depth of insertion, withdraw the probe and measure the distance directly against the scales (refer to diagram). A word of warning about probing for hemipenes: it is very easy to injure the reptile if you push too hard or if your equipment is unsuitable. It is therefore recommended that you get assistance from someone with experience in the technique for your first attempts.

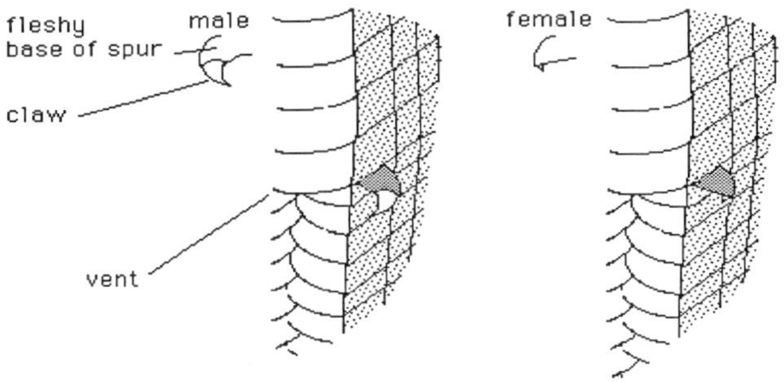

Figure 16 Adult diamond pythons and carpet pythons are not easily sexed by probing, but can usually be distinguished upon inspection of the paired 'spurs' - situated to either side of the vent. Males have longer claws extending from the spurs than do females of similar size.

This technique should not be used for adult diamond/carpet pythons. Females have paired ducts positioned lengthwise within the base of the tail, simulating hemipenes. These ducts are readily entered and ruptured with the probe when they are often mistaken for hemipenes. A safer technique for sexing adult diamond pythons and carpet pythons is to compare the length of the 'spurs' situated at the base of the tail of both sexes - one on either side of the vent. In males, the claw at the end of this spur will be far longer than in a female of equal length (see figure 16). In fact, the claw hardly protrudes from the spur in females under 180 cm length, while it is quite pronounced in males of greater length than 80 cm.

Separating and Reintroducing Sexual Pairs of Reptiles

It is generally agreed that there is value in the practice of keeping male and female reptiles (particularly lizards and snakes) separate during all but the month(s) in which successful matings would naturally occur in the wild. Familiarity through long periods together tends to lead to disinterest in mating, or excessive mating may occur at times when the female is not capable of conceiving. When she is finally in the reproductive condition, you may well find that the male is no longer interested. A popular practice is to put males and females together for a trial period of a couple of days to see if any mating occurs. These short periods of cohabitation should be allowed every two weeks or so during the known mating season of the species. Of course there are some species that seem to reproduce no matter how your manipulate or arrange them and both sexes of these species can generally be kept together year round and still be expected to breed regularly. Examples of these are Children's pythons, water pythons, pink-tongued skinks, Cunningham's skinks, and probably many more.

The Use of Multiple Males in Breeding

Not all reproductively capable reptiles are compatible with each other, and merely placing a male and female together at the correct time of year will not ensure reproductive success. Sometimes it is best to house several specimens together during the known breeding period or for short periods as described above. An additional advantage in introducing more than one male at a time to a potentially receptive female is that often the competition between males appears to lead to more aggressive mating behaviour.

Courtship displays and activities vary widely among reptiles. In many species of lizards, head bobbing, arm waving, and other means of communicating male aggression may occur. This may be followed by bouts of fighting. In some species these encounters can be injurious and potentially fatal if the submissive males have no effective means of escape. In confined circumstances, you should watch closely to ensure that a dominant male does not inflict damage upon any of its weaker cage mates. Amongst the snakes, male aggression is usually confined to entwined wrestling bouts, during which biting is unusual. Carpet pythons and black-headed pythons are an exception to this, and males have been known to inflict deep wounds to one another by dragging the upper row of their teeth across the dorsal surface of their rivals. Male tortoises may fight amongst themselves during the breeding season, and they should be separated if these skirmishes occur continually.

'Cooling Down' Reptiles Prior to Breeding

Most Australian reptiles are seasonal in their reproductive efforts. It is well known amongst reptile breeders that many species rely to some extent upon environmental cues to stimulate reproductive activity. In the absence of the necessary environmental cues, often the case in captive situations where artificial light and heat sources remain constant year round, some reptiles will not produce eggs or young despite repeated matings.

Publications by reptile breeders suggest that if temperate zone reptiles are to be bred indoors, it is necessary to imitate the major climatic trends that naturally occur outdoors. Most significant of these conditions is the provision of a 'cooling down' period during the winter months, followed by a spring warming. The timing of the cooling and warming periods should reflect the natural cycle of the seasons - so that cooling should begin during May or June, and warming in August or September. During the winter months many species will not feed, and may become very lethargic or even enter a state of hibernation.

Snakes that are cooled down generally remain fairly active, but typically show no interest in food. However they do drink water, and should be given an opportunity to drink at least once a week. Some keepers simply stop providing heat to their reptiles during the winter months. This can result in conditions that are unduly cold since it removes the opportunity for reptiles to sun bask, an activity possible under natural conditions for most Australian species. A more sensible strategy is to direct a basking light (see heating section of this booklet) onto an area of the cage floor for five or six hours a day, but then to turn the light and all other heat sources off during the remainder of the day and night.

Specimens maintained in 'winter' conditions should be monitored and checked regularly for signs of disease and their enclosures kept clean. If excessive weight loss occurs in these reptiles, or if signs of skin or respiratory complaints are detected, the specimens should be brought back into 'summer' conditions and treated. Breeding using these individuals will most likely have to be postponed for a year.

Some of the species for which a cooling down period is recommended as a prerequisite to breeding are identified in the individual care sheets. The sort of temperature regime regarded as suitable for a particular species may be quite different to that recommended for another. For example, certain northern species do well and breed when kept at a constant warm temperature twelve months a year, while southern alpine reptiles might need to be cooled to a point where hibernation occurs for a period of the year in order to stimulate breeding. Under no circumstances should any reptile that is in sub-optimal health or underweight be cooled down.

Mating Habits

The mating habits of reptiles are highly variable. Typically, however, during the 'mating season', the males will court available females and perhaps combat any rival males that are present. In most lizards, the successful male partner grasps the female in his jaws, usually at the back of the neck, and brings his cloaca into contact with that of the female in order to facilitate mating. Male snakes often initiate mating behaviour by rubbing their chin along the female's back and the top of her head, while simultaneously attempting to restrain her movement with loops of his body. Once the male's cloaca has been manoeuvred to lie adjacent to the female's, mating proceeds. The period of contact can be as brief as a few minutes for lizards, or may last as long as 24 hours in snakes. Tortoises mate in the water; the male mounting the female dorsally from behind. Mating may occur several times over a few days. Females of reptile species can store sperm from a single mating for several months and subsequently produce fertile eggs.

Care of Gravid (Pregnant) Reptiles

Female reptiles are said to be 'gravid' prior to depositing their clutch of eggs or giving birth to their brood of live young. Following fertilization, female lizards and snakes will appear to grow progressively heavier, despite the fact that they may not be feeding. Most will in fact refuse food during the later stages of pregnancy, prior to egg laying or giving birth. Water should be provided regularly, but as the time of birth or laying approaches, the water bowl should be removed from the cage when unattended, since the eggs or young may be deposited in the bowl and drown. Keep in mind also that newly born lizards and snakes can be quite tiny, and may easily escape out of any small gaps in the cage. Also, gravid live-bearing reptiles should be housed alone to avoid the possibility of cage mates eating the young.

If your reptile is a diurnal species that is kept indoors, or if its diet has been less than satisfactory during previous months, it is a good idea to supplement its diet with a Vitamin D/calcium preparation during the early stages of its pregnancy before it ceases feeding (see section on vitamin supplementation).

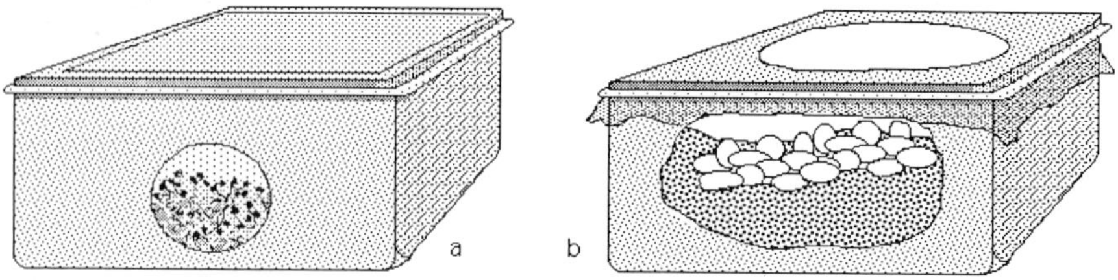

Figure 17 (*a*) Egg-laying box for snakes or lizards. Slightly dampened peat moss or vermiculite can be seen through the access hole in the front. (*b*) Cutaway view through the front of an incubating container showing a clutch of eggs on a mixture of vermiculite and water (1:1 by weight or drier). A large hole in the lid, and a sheet of clear plastic wrap sandwiched beneath the lid, allow viewing during incubation. Once sealed, there is no reason, save spoilage of eggs, to open the container until hatching.

The thermal requirements of gravid females are usually fairly high. Tortoises and lizards kept outdoors will be able to bask and maintain suitably high body temperatures. For lizards and snakes maintained indoors, however, it is best to provide a reasonably broad range of temperatures across the cage to ensure that the gravid female has every opportunity of maintaining her body temperature at an optimal level. Where practical, gravid lizards kept outdoors should be moved indoors. Egg laying species should be provided with a suitable egg deposition site that is dark, warm and humid. Conditions in the enclosure should be such as to protect the eggs from desiccation, abrasion, and excessive movement. A plastic 'Tupperware' - type box or a plastic ice cream container will make a good laying site. This should be half filled with slightly dampened sphagnum moss or vermiculite, and have at least one reasonably large access hole (see diagram). The box can be opened occasionally to see if the eggs have been laid and will help to ensure the eggs do not desiccate before being discovered.

Large lizard species such as water dragons and goannas should be kept outdoors when gravid. You will have to watch them closely for signs of nest digging (or sudden weight loss) in order to retrieve the eggs for artificial incubation.

Incubation of Reptile Eggs

The successful incubation of lizard and snake eggs has become a very straightforward process in recent years, and in the event of fertile, white-shelled eggs, can nearly be assured. Infertile or dead eggs are usually a cream to yellowish colour upon deposition (though they may also be indistinguishable from live, fertile ones) and soon take on a slimy appearance and become solidified within. 'Windows' or uncalcified patches may be present upon the shells of good or bad eggs, and these may disappear during the course of incubation.

Incubation Containers and Media

The incubation medium of choice is vermiculite (available at nurseries and gardening supply shops) which is usually mixed with an equivalent *weight* of water. Some hobbyists,the author included, believe that a drier mixture is more suitable for the eggs of most species, mixing the vermiculite and water at a ratio of 6:4 or 7:3 (for arid-zone species) respectively by weight. Make sure that the vermiculite is in the dry form before mixing. The mixture should be put into a clean plastic ice cream container or equivalent (four litre size or bigger). A window can be cut in the ice cream container lid, and a sheet of 'glad wrap' plastic sandwiched between the lid and the container to allow observation of the eggs during incubation without opening the container (see figure 17). No appreciable amount of water will escape from the medium during the incubation process, so it is unlikely that there will be a need to add any. Once sealed, the incubation container must be placed in a position of constant warmth. A thermometer should be kept within the incubation container in such a position as to be visible through the 'window' in its lid.

Incubators

The cheapest, though not the most reliable heating system to warm the incubation container involves the use of a styrofoam esky or similar shipping container (with a tight fitting lid) placed on a thermostatically controlled medical or home heating pad. The heating pad should be turned to the lowest setting. It may take a day or more for the chest and the incubating container within to equilibrate to an even temperature. If more warmth is required, the pad's setting may be slowly increased until the desired temperature is achieved in the vermiculite. The best way to incubate reptile eggs is to build an incubator such as the one in the diagram provided. If you are expecting a clutch of eggs from one of your reptiles, you should have the incubator and incubating material prepared in advance and the temperature within stabilised at the desired level. Depending on the species involved, temperatures between 26° and 30°C are acceptable, with a constant 28°C being a safe choice for most.

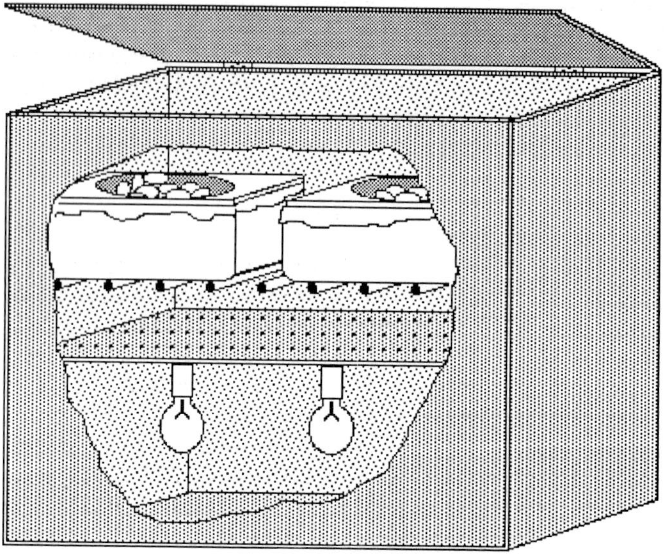

Figure 18 Cutaway view through the front of a home made reptile egg incubator. Heat produced by light globes in the lower chamber passes upwards through a sheet of pegboard to the upper chamber where the egg containers rest upon wooden dowels. Termperature is controlled by a thermostat mounted in the upper chamber.

Most lizard and snake eggs are oval and somewhat soft and leathery when laid - the exception being some gecko eggs, which are round, hard, and brittle. You will notice that leathery-shelled eggs will expand during development as they take in water. Eggs touching at laying have a tendency to stick to each other. This is nothing to worry about, and they shouldn't be forced apart as this may cause tears in the shell. The eggs can be placed as a single mass or separately within the incubating medium and it doesn't appear to matter whether the eggs are partially or completely buried in the vermiculite. However, it is useful to allow some part of the egg to project so that they can be inspected without disturbance. Make sure that the up/down orientation of the eggs as laid is not altered when they are moved to the incubating container. Together, the eggs and vermiculite mix should only half fill the incubation container, the rest of the space being taken by air. The volume of eggs to volume of vermiculite in container should not exceed 1:4. Eggs that develop a yellow, slimy appearance should be removed (if separate). Avoid handling or turning good eggs once the incubation process has begun.

Egg incubation times and gestation periods for live bearers will vary somewhat between reptile species and to a degree within a single species. In general, the warmer an incubating egg or pregnant female is kept, the faster development occurs. However, it is important to maintain temperatures within the recommended 'safe zones' for any given species, to avoid deformity or death of the embryos.

Care of Hatchling and Newly Born Reptiles

When a lizard or snake is ready to hatch it will use its 'egg-tooth', a tiny, sharp projection on the tip of the snout, to cut one or more slits in the egg. You will notice droplets of clear fluid oozing from these slits prior to hatching. Tortoises are able to use their snout and claws to assist in hatching. It is best to allow the hatchlings to emerge from their egg shells unassisted and undisturbed, even though this might take several days after the slits appear, as remaining yolk sacs are absorbed. Eggs that fail to show signs of hatching after the expected incubation period should be left in the incubator, particularly when none of the eggs in the clutch have hatched. As long as the eggs are not slimy or solidified, they are possibly still viable and may hatch in due course.

The only exception to this lack of intervention is when one or two eggs in a clutch remain unhatched and show no sign of slitting. Sometimes, in the case of lizards or snakes, the egg-tooth is not properly formed and the baby reptile may be trapped within the egg. In this case, you can use a razor blade to delicately etch a slit of appropriate length into the egg.

The situation with live-born baby reptiles is slightly different. The young are usually born encased in a transparent membrane. You should not attempt to open these sacs to free the young. They will release themselves when ready, and can survive in the sac for several days, drawing additional nutrients from any remaining yolk mass. You may, however, carefully move the egg sacs into a warm, damp mixture of either sphagnum moss or vermiculite.

The umbilical slit that is present in all newly hatched or born reptiles will soon heal, and the 'egg-tooth' will disappear within a couple of days. For a short period after birth or hatching, the young are susceptible to desiccation and should be moved into conditions of increased humidity, with a constant source of water. Make sure that they can get to, and if need be, climb out of the water receptacle provided.

Because hatchlings or newly born reptiles often have a fair amount of food reserve, they may refuse to feed for up to two weeks. If they do not feed within four to six weeks, and you have checked to make sure that they have correct captive conditions and that the proper foods have been offered, active measures such as force-feeding (see section on involuntary-feeding) may have to be adopted.

Newly born or hatched diurnal lizards and tortoises need access to sufficient ultraviolet light, preferably sunshine, as soon as practical. Sunlight should be direct and not filtered through glass or plastic.

Dispensing of Your Excess Captive-bred Progeny

Newly born or hatched reptiles generally require individual care. Ideally, each young specimen should be provided with its own separate caging facility until it begins feeding. This is a good reason to dispense with your captive progeny fairly early if this is your intention.

Although generally regarded as a noble gesture, it is a biologically unsound idea to release captive bred young reptiles into the wild randomly. Introducing individuals into the wild increases the competition for available food reserves. This means that either members of the same species will be displaced because of territoriality or direct competition, or that the increased pressure on the shared food resources could eventually cause starvation of inhabitants. The other possibility, which is in fact more probable, is that the vast majority of released young will very soon be eaten by the natural predators of the area. This is the usual destiny of the majority of wild newly born or hatched reptiles.

The only instance where the release of captive-bred specimens into the wild is beneficial is in the situation where a population of the species had been drastically reduced in the area, and the cause of that reduction had been remedied. This is a combination of events that is rarely encountered in these times of active urban and agricultural sprawl and development.

A far more sensible solution to the problem of disposing of your captive progeny is to distribute them to other keen reptile keepers that would like to have them. If you have trouble locating such people, contact your herpetological association as soon as your young are hatched or born, and they will assist you in finding good homes for your reptiles.

Captive bred animals in general are free of parasitic diseases, and are ideal for transfer to other reptile keepers. By distributing your excess young you will not only effectively reduce the need for others to be collected from the wild, but will also lessen the burden on yourself for providing proper care for a large number of small animals. The ideal time to transfer the young to another keeper is as soon as it has digested its first meal, or at an age of two to three weeks - whichever comes first. If possible, sex the young (or seek assistance from an experienced person) so that you can provide sexual pairs to other interested hobbyists.

You can contribute to the pool of knowledge relating to reptile propagation by writing up the results of your breeding accomplishments and/or difficulties and publish them in newsletters and journals produced by herpetological organizations. This is particularly important where you have experimented with new strategies or techniques and have been successful.

Care Sheets: Maintenance of Individual Species

ROBUST VELVET GECKO *Oedura robusta*
TRYON'S VELVET GECKO *Oedura tryoni*
BROAD-TAILED GECKO *Phyllurus platurus*
(and other large arboreal geckos)

robust velvet gecko

Tyron's velvet gecko

broad-tailed gecko

HABITS: These species are avid climbers, foraging at night on rock faces and tree trunks for insects and other invertebrates. During the day they rest in the hollows or crevices of dead trees, beneath loose bark, or in rock crevices or caves.

HOUSING: An adult pair of any of these species will need an indoor terrarium at least 40cm long. Cage decoration should include a suitable separate hiding place for each specimen. These can consist of crevices created with small flat rocks arranged in a pattern resistant to collapse, pieces of bark to crawl beneath, or even small cardboard boxes with access holes cut out. Hiding places should be placed in warm positions in the cage.

To stimulate the nocturnal activity of specimens and to improve your ability to watch them, a red or blue globe should be installed to warm the cage by night. The cage can be warmed by day using the same globe, or by installing a second light fitting for a normal white light globe of appropriate wattage to be turned on during daylight hours as needed. A gradual cline of temperatures from one end of the cage to the other is preferred, and can be promoted by placing the primary heat source towards one end of the cage. The temperature in the middle of the cage should be maintained between 24 and 28°C.

Besides the usual continuous provision of water from a small bowl, the cage interior should be sprayed with a fine mist from time to time as geckos often prefer to obtain water by licking up droplets.

FEEDING: A variety of insects such as mealworms, crickets, locusts, cockroaches and moths will be accepted by the geckos. Large specimens may accept pink mice. Food should be offered every two to three days.

CAPTIVE BEHAVIOUR: Although shy in the wild, these geckos will adjust readily to captivity, though not to handling, which should be kept to a minimum. Long-term captives may emerge and take food during the day.

BREEDING: Adult males can be recognized by the presence of a pair of bulbous protrusions positioned just behind the vent. These represent the inverted hemipenes. All Australian geckos are egg layers, with two eggs per clutch.

THICK-TAILED GECKO ***Underwoodisaurus milii***
(and other large ground-dwelling geckos)

thick-tailed gecko

HABITS: These are ground dwelling, nocturnal species that shelter by day beneath large objects or within burrows . At night they can be found foraging on the ground.

HOUSING: A pair of these geckos may be housed in an indoor terrarium at least 40cm long. Cage decoration should include suitable hiding places (such as appropriately sized and shaped pieces of bark or flat rocks) for day time shelter situated in a warm area of the cage.

To stimulate the nocturnal activity of specimens and allow you to watch them, a red or blue globe should be installed to warm the cage by night. The cage can be warmed by day using the same globe, or by installing a second light fitting for a normal white light globe of appropriate wattage to be turned on during daylight hours as needed. A gradual cline of temperatures from one end of the cage to the other is preferred, and can be promoted by placing the primary heat source towards one end of the cage. The temperature in the middle of the cage should be maintained between 24 and 28°C.

Besides the usual continuous provision of water from a small bowl, the cage interior should be sprayed with a fine mist from time to time as geckos often prefer to obtain water by licking up droplets.

FEEDING: A variety of insects such as mealworms, crickets, locusts, cockroaches and moths will be accepted by the geckos. Large specimens may accept pink mice. Food should be offered every two to three days.

CAPTIVE BEHAVIOUR: Although shy in the wild, these geckos will adjust readily to captivity, though not to handling, which should be kept to a minimum.

BREEDING: Adult males can be recognized by the presence of a pair of bulbous protrusions positioned just behind the vent. These represent the inverted hemipenes. All are egg layers, with two eggs per clutch.

BURTON'S LEGLESS LIZARD *Lialis burtonis*
HOODED SCALY-FOOT *Pygopus nigriceps*

Burton's legless lizard

hooded scaly-foot

HABITS: Both species are terrestrial and may be active by either day or night, depending upon the weather. They are most commonly encountered crossing roads during the very early evening hours. When inactive, they may shelter in leaf litter and spinifex grass clumps, as well as beneath rocks, logs, etc. Burton's legless lizards feed almost entirely upon other lizards, while hooded scaly-foot lizards prey mainly upon spiders and other invertebrates.

HOUSING: A pair of either species should be kept in an indoor terrarium at least 50cm long. Cage decoration should include a suitable hiding place (such as appropriately sized and shaped pieces of bark, etc.) situated in a warm area of the cage.

To stimulate the nocturnal activity of specimens and to improve your ability to watch them, a red or blue globe should be installed to warm the cage by night. The cage can be warmed by day using the same globe, or by installing a second light fitting for a normal white light globe of appropriate wattage to be turned on during daylight hours as needed. A gradual cline of temperatures from one end of the cage to the other is preferred, and can be promoted by placing the primary heat source towards one end of the cage. The temperature in the middle of the cage should be maintained between 24 and 28°C.

FEEDING: Burton's legless lizards should be fed lizards, offered once a week, while hooded scaly-foot lizards will accept a wide variety of insects and spiders, offered every two to three days.

CAPTIVE BEHAVIOUR: Both species adjust readily to captivity, though not to handling, and should be handled as little as possible. The long tail is readily shed with rough handling. Burton's legless lizards will often lay concealed beneath leaf litter, a dried grass tussock, or whatever other cover you may choose to include in the terrarium, with just the end of the snout protruding, waiting for the opportunity to ambush any passing food item. When it does detect food, it may lunge out at remarkable speed.

BREEDING: It is very difficult to determine the sex of legless lizards. Burton's legless lizards mate in the spring and two eggs are laid, usually in December or January.

COMMON SCALY-FOOT *Pygopus lepidopodus*
(and other non-burrowing flap-footed lizards)

common scaly-foot

HABITS: The common scaly-foot feeds mainly on spiders in the wild, though it also takes various insects on occasion. Other, smaller species of flap-footed lizards feed largely on insects. When inactive, they shelter in leaf litter. The long tail of these species is very prone to loss when captured.

HOUSING: Flap-footed lizards should be kept indoors. Cage decoration should simulate their natural habitat, with a soil substrate and a large rock or flat timber beneath which they can shelter. Care should be taken that the shelter site is stable and cannot settle to crush the lizard. An indoor enclosure for one or two adult common scaly-foot lizards should be at least 50cm long, with a warmed area near the shelter site. A basking position is necessary, and is best provided with a reflector globe.

FEEDING: Captive specimens will readily take small insects, spiders, and soft fruit. They should be fed every two to three days.

CAPTIVE BEHAVIOUR: Captive specimens soon become used to human presence and will remain active when disturbed, although they should be handled as little as possible.

BREEDING: It is difficult to determine the sex of common scaly-foot lizards. Females lay a clutch of two eggs in the spring, following winter cooling. Young individuals should be housed separately from the adults as small specimens may be eaten.

LACE MONITOR *Varanus varius*

lace monitor

CAUTION: Lace monitors have sharp, shearing-type teeth and strong claws to allow them to climb and feed on carrion. Therefore, care should be used when handling them initially, as their bite and scratch can cause deep lacerations that require medical attention.

HABITS: The lace monitor is a diurnal, somewhat arboreal species that may be encountered foraging either on the ground or in trees. It normally preys upon any small animal which can be overpowered, but also feeds extensively upon carrion.

HOUSING: If this species occurs naturally in the area, it may be kept in a large (minimum ground dimensions: 3m x 3m) outdoor enclosure with smooth walls extending at least 60cm below the ground, and 180cm above the ground. Alternatively an aviary type enclosure of equal floor dimensions will suit. Either type of enclosure should be situated in a position where it will receive more than 8 hours of sun a day, but should also provide some shaded and dry areas.

FEEDING: Lace monitors readily feed on meat, mice, rats, young chickens, etc. They should be fed once every week or fortnight. Care should be taken to avoid overfeeding, which readily occurs in captivity due to the lack of activity.

CAPTIVE BEHAVIOUR: Captive lace monitors spend a great deal of time sun basking on logs or whatever is available. They soon become accustomed to human presence.

BREEDING: Sex determination for this species is problematic, but may be accomplished by X-raying. Consult your herpetological association for information on sexing goannas. An egg layer, the female lace monitor lays her clutch of approximately 8 eggs in the autumn or spring.

SAND GOANNA (OR GOULD'S MONITOR) ***Varanus gouldii***
(and other large goannas)

sand goanna

HABITS: This is a diurnal, sun-loving species, commonly seen basking on roads, amongst rocks or on branches of trees. It is both an excellent climber and an efficient burrower, and when frightened will seek refuge in either trees or available burrows. Likewise, foraging is done at ground level in animal burrows, and in trees. Sand goannas normally eat any small animal which can be overpowered, as well as carrion. Sand goannas have sharp, shearing-type teeth and impressive claws to allow them to feed on carrion and to climb. Care should be used when handling them initially, as their bite and scratch can cause deep lacerations that require medical attention.

HOUSING: If these species occur naturally in the area, they may be kept in a large outdoor enclosure with smooth walls extending at least 100cm below the ground, and 180cm above the ground. Such an enclosure should be situated in a position where it will receive more than 8 hours of sun a day, but should also provide some shaded and dry areas. Sand goannas may also be kept outdoors in eastern coastal areas, but special attention must be paid to design to ensure dry conditions within (eg. transparent fibreglass roofing can be provided).

To house a pair of small adults (snout to vent length to 40cm) indoors you will need a terrarium at least 130cm long and 60cm wide. Larger specimens will require more room. Cage decoration should include suitable hiding places for use at night.

Heating should be provided at least in part by inclusion of a reflector globe within the cage, positioned in such a manner as to allow basking. The cage temperature (away from the basking site) should be kept between 22°C (night-time low) and 30°C. The temperature at the basking site will of course be higher.

FEEDING: Goannas may be fed an assortment of insects as well as mice or young rats. Goannas are active lizards and adults should be fed once a week.

CAPTIVE BEHAVIOUR: Initially sand goannas will be nervous, but if given ample quiet and privacy, will adjust readily to captivity. In time some individuals become amenable to handling.

BREEDING: Sex determination for goannas is problematic, but may be accomplished by X-raying. Consult your herpetological association for information on sexing goannas. All goannas are egg-layers. The female sand goanna lays her clutch of between 5 and 8 eggs in the spring or summer.

SPINY-TAILED MONITOR *Varanus acanthurus*
CENTRAL NETTED DRAGON *Amphibolurus nuchalis*

spiny-tailed monitor

central netted dragon

HABITS: Both species are commonly seen basking amongst rocks beneath which burrows are excavated, though crevices are also inhabited by spiny-tailed monitors. Both feed on insects.

HOUSING: To house a pair of adults of either species you will need a terrarium at least 90cm long and 40cm wide. Cage decoration should include suitable hiding places for use at night.

Heating should be provided at least in part by inclusion of an ultraviolet light globe (eg Gro-Lite brand), positioned in such a manner as to allow basking. The cage temperature (away from the basking site) should be kept between 22°C (night-time low) and 30°C. The temperature at the basking site will of course be substantially higher.

FEEDING: These species may be fed an assortment of insects as well as very small mice and canned pet food. Feeding should take place every two to three days. Meals should occasionally be coated with a calcium/vitamin D powder.

CAPTIVE BEHAVIOUR: Initially these species will be nervous, but if given ample quiet and privacy, will adjust readily to captivity. In time some individuals become amenable to handling.

BREEDING: Adult male central netted dragons can be recognized by the presence of enlarged femoral and preanal pores. Sex determination for the monitors is problematic, but may be accomplished by X-raying. Consult your herpetological association for information on sexing goannas.

JACKY LIZARD *Amphibolurus muricatus*
(and other moderate-sized semi-arboreal dragons)

jacky lizard

HABITS: Jacky lizards are terrestrial, semi-arboreal, sun-loving species commonly seen basking beside roads or on fence posts and logs. In the natural state they eat a wide variety of insects and also some vegetation such as yellow daisies.

HOUSING: If the species considered occurs naturally in the area, it may be kept in a large outdoor enclosure with smooth walls extending at least 60cm below the ground, and 150cm above the ground. Such an enclosure should be situated in a position where it will receive more than 8 hours of sun a day, but should also provide some shaded and dry areas.

If being housed indoors, one or two adults will need an indoor terrarium at least 100cm long. Cage decoration should include branches for climbing and suitable hiding places, such as deep leaf litter for use at night. Heating should be provided at least in part by inclusion of an ultraviolet light globe (eg Gro-Lite brand), positioned in such a manner as to allow basking. The cage temperature (away from the basking site) should be kept between 25°C (night-time low) and 32°C, although the temperature in the basking site should be somewhat higher.

In addition to a continuous source of fresh water, the interior of the cage should be occasionally sprayed with a fine mist as dragons often prefer to obtain water by licking up droplets.

FEEDING: These species may be fed a wide assortment of insects as well as young mice, minced meat (which should be supplemented with calcium/vitamin D powder), bananas and dandelion flowers. All are active lizards, and adults and juveniles should be fed every two to three days. Meals should occasionally be coated with a calcium/vitamin D powder.

CAPTIVE BEHAVIOUR: Initially these dragons will be nervous, but if given ample quiet and privacy will adjust readily to captivity.

BREEDING: Adult males can be recognized by the presence of enlarged femoral and preanal pores. All egg layers; the female jacky lizard lays an annual clutch of six to eight eggs, usually in the late spring – early summer.

WATER DRAGON *Physignathus lesueurii*

water dragon

HABITS: This is a semi-aquatic, semi-arboreal, sun-loving species commonly seen basking amongst rocks or on branches overhanging the water. In the natural state water dragons eat a wide variety of insects and crustaceans, as well as frogs, reptiles and small mammals. Berries and fruits are also eaten upon occasion.

HOUSING: If this species occurs naturally in the area, it may be kept in a large outdoor enclosure with smooth walls extending at least 60cm below the ground, and 150cm above the ground. Such an enclosure should be situated in a position where it will receive more than 8 hours of sun a day, but should also provide some shaded and dry areas.

If being housed indoors, an adult pair of water dragons will need to be kept in a terrarium at least 130cm long. Cage decoration should include branches for climbing, and suitable hiding places. Heating should be provided at least in part by inclusion of an ultraviolet light globe (eg Gro-Lite brand), positioned in such a manner as to allow basking. The cage temperature (away from the basking site) should be kept between 25°C (night-time low) and 32°C, although the temperature at the basking site must be somewhat higher. A continuous supply of clean water should be available in a container large enough to allow both specimens to completely submerge themselves without causing the water to spill over.

FEEDING: Water dragons may be fed an assortment of insects such as mealworms, crickets, locusts, cockroaches and moths, as well as young mice or rats, worms, snails (collected in unbaited areas only), minced meat (which should be supplemented with calcium/vitamin D powder), and bananas. This is an active species; adults should be fed once a week, and juveniles every two or three days. Meals should occasionally be coated with a calcium/vitamin D powder.

CAPTIVE BEHAVIOUR: Initially water dragons will be nervous, but if given ample quiet and privacy will adjust readily to captivity. In time most individuals become amenable to handling.

BREEDING: Adult males can often be recognized by their greater size and much brighter colouration. An egg layer, the female water dragon lays an annual clutch of 6-15 eggs, usually in the Late spring or summer.

BEARDED DRAGONS genus *Pogona*

western bearded dragon

HABITS: Bearded dragons are terrestrial, sun-loving species commonly seen basking on or beside roads or on fence posts and logs. In the natural state they eat a wide variety of insects and plant matter. Large specimens may also eat small lizards and snakes.

HOUSING: If the species occurs naturally in the area, it may be kept in a large outdoor enclosure with smooth walls extending at least 60cm below the ground, and 150cm above the ground. Such an enclosure should be situated in a position where it will receive more than 8 hours of sun a day, but should also provide some shaded and dry areas. If being kept indoors, one or two adults will require a terrarium of at least 100cm length. Cage decoration should include branches for climbing, and suitable hiding places, though bearded dragons will usually sleep in an open position at night.

Heating should be provided at least in part by inclusion of an ultraviolet light globe (eg Gro-Lite brand), positioned in such a manner as to allow basking. The cage temperature (away from the basking site), should be kept between 25°C (night-time low) and 32°C, though the temperature at the basking site will be somewhat higher.

In addition to a continuous source of fresh water, the inside of the cage should be occasionally sprayed with a fine mist, as dragons often prefer to obtain water by licking up droplets.

FEEDING: Bearded dragons may be fed an assortment of insects such as mealworms, crickets, locusts, cockroaches and moths, as well as young mice or rats, worms, snails (collected from snailbait-free areas only), minced meat (which should be supplemented with calcium/vitamin D powder), bananas, chopped lettuce, dandelion flowers and other edible weeds, such as clover, milk thistle, etc. Bearded dragons are active lizards and adults should be fed at least once a week, and juveniles every two to three days. Meals should occasionally be coated with a calcium/vitamin D powder.

CAPTIVE BEHAVIOUR: Initially specimens will be nervous, but if given ample quiet and privacy will adjust readily to captivity. In time, most individuals become amenable to handling and will display, with head bobbing and arm waving.

BREEDING: Adult males can be recognized by enlarged femoral and preanal pores. All bearded dragons are egg layers; the female western bearded dragon lays an annual clutch of 8 to 16 eggs in the late spring or summer.

TREE SKINK ***Egernia striolata***
WHITE'S SKINK ***E. whitii***
(and other medium-sized *Egernia* species)

tree skink

White's skink

HABITS: White's skinks generally live in small colonies or family groups associated with rock outcrops or logs, under which they dig shallow burrows or cavities. Some populations inhabit rock crevices. Tree skinks are arboreal or rock-dwelling, sheltering in crevices under bark and in dead timber, or in rock crevices, and sunning by day close to shelter. Generally only one or two adults inhabit a single tree, although young tree skinks may stay with the parents for up to a year. All are sun-loving skinks often seen basking close to retreats to which they flee if disturbed. All Egernia species are omnivorous, feeding largely on insects but also taking some vegetable matter such as small berries.

HOUSING: Although tree skinks and White's skinks may be kept outdoors in warm, dry climates, the enclosure must be smooth-sided as they are good climbers. Indoor enclosures allow much better visibility and care. Cage decoration should include a crevice, such as a flat rock on rock or a log with loose bark for each lizard to shelter in. Care should be taken that the shelter site is stable. An indoor enclosure for a pair of adult tree skinks or White's skinks should be at least 50cm long with a warmed area near the shelter site, preferably provided by an incandescent globe.

FEEDING: Both species are omnivorous in captivity, eating commercial canned pet foods (stay clear of fish-types), meat, and some vegetable material, but preferring insects such as mealworms, crickets, locusts, cockroaches and moths.

CAPTIVE BEHAVIOUR: Although shy in the wild and in outdoor enclosures, captive specimens kept indoors rapidly become tame, and will remain active when disturbed. However they rarely adjust to handling, and should therefore be handled as little as possible. Disturbance of the shelter site should be kept to a minimum.

BREEDING: Both species produce one to five young (usually two or three) in early summer, generally January. Mating occurs in spring, following winter chilling. New-born young are about 8cm long, much of which is tail, and readily feed on the same diet as adults. Young individuals should be housed separately from the adults.

LAND MULLET *Egernia major*
MAJOR SKINK *E. frerei*
(and other large *Egernia* species)

land mullet

major skink

HABITS: Both land mullets and Major skinks are primarily ground-dwelling, taking refuge within burrows and fallen hollow logs. During the early morning and mid-afternoon, they alternately bask in sunny patches between episodes of foraging for a wide range of foods, both plant and animal. Most commonly encountered at the edges of thick vegetation (usually either rainforest or wet sclerophyll forest), where they sun bask.

HOUSING: Where the climate suits (ie districts as warm or warmer than areas where the species naturally occurs), these species may be kept in a large outdoor enclosure with smooth walls extending at least 50 cm below ground and 100 cm above ground. Such an enclosure should be situated in a position that will receive more than 8 hours of sun a day, but should also provide some shaded and dry areas. Land mullets and Major skinks may also be kept indoors in smaller enclosures. A cage for one or two adults must be at least 100cm long and provide a basking area. Shelter in the form of a low flat box, leaf litter or loose grass should also be provided.

FEEDING: These large skinks will feed readily on a wide variety of food, including soft fruits, insects, small mice, and commercial tinned pet food (non-fish varieties). The diet should not be confined exclusively to a single food type for long periods. Snails and slugs will also be eaten, although these should only be collected from areas without snail baits. A once weekly feeding schedule is suitable for adults, though juveniles should be fed every three to four days.

CAPTIVE BEHAVIOUR: Although often timid and liable to thrash about wildly and bite when initially collected, these skinks will rapidly become settled. However, handling should be kept to a minimum.

BREEDING: Both species are live bearers. Land mullets give birth to up to six live young in spring/early summer. Mating occurs in spring. New-born young of both species readily feed on the same diet as adults. Young individuals should be housed separately from the adults.

NARROW-BANDED SAND SWIMMER *Eremiascincus fasciolatus*
BROAD-BANDED SAND SWIMMER *E. richardsoni*

broad-banded sand swimmer

HABITS: As the name suggests, sand swimmers spend most of the daylight hours below the surface in loose sand, often with just the nose protruding. After dark on warm nights they emerge from the sand to forage on the surface for insects, spiders or small reptiles. They also ambush their prey by lying motionlessly just beneath the surface of the sand in anticipation of approaching food. When disturbed, these skinks rapidly 'swim' into the loose sand to escape. Sand swimmers are most commonly found active on the surface of dunes or sandplains at night, or in the loose sand below vegetation by day.

HOUSING: Sand swimmers should be kept indoors. An indoor enclosure for one or two adults should be at least 50cm long. Cage decoration should simulate their natural habitat, with a sandy substrate, preferably obtained from the area where they were captured, and dry vegetation, bark or small rocks to provide extra shelter.

To stimulate the nocturnal activity of specimens and to improve your ability to watch them, a red or blue globe should be installed to warm the cage by night. The cage can be warmed by day using the same globe, or by installing a second light fitting for a normal white light globe of appropriate wattage to be turned on during daylight hours as needed. A gradual cline of temperatures from one end of the cage to the other is preferred, and can be promoted by placing the primary heat source towards one end of the cage. As these species are desert inhabitants preferring high temperatures, the sand surface temperature below the heat lamp should be kept at 28-32°C, although it may be cooler elsewhere in the cage. Water should always be available in a non-spillable container.

FEEDING: Sand swimmers should be fed on live food, including insects, small spiders and small lizards, which they will ambush from below the sand. A well fed specimen will store fat in the tail, which in cases of obesity may become very distended and of greater diameter than the body.

CAPTIVE BEHAVIOUR: Specimens rapidly become used to human presence and will remain active when disturbed. In time they become amenable to handling.

BREEDING: Two to seven eggs are laid in late spring. The young are about 5cm long upon hatching, most of which is tail. They should be housed separately from the adults, which readily eat such small lizards.

MOUNTAIN WATER SKINK *Eulamprus heatwolei*
GOLDEN WATER SKINK *E. quoyii*

(and other water skinks)

mountain water skink

golden water skink

HABITS: Water skinks feed upon a wide range of invertebrates, as well as berries and fruit in the wild. Individuals are easily approached as they sun bask and can occasionally be observed giving chase to potential prey.

HOUSING: Both species may be kept outdoors in areas which they naturally inhabit, although they are more visible and more easily cared for if kept indoors. Cage decoration should simulate their natural habitat, with a soil substrate and a large rock or flat timber below which they can shelter. Care should be taken that the shelter site is stable and cannot settle to crush the lizard. An indoor enclosure for an adult pair should be at least 50cm long with a warmed area near the shelter site. A basking position is necessary and this is best provided with a reflector globe.

The water bowl should be large enough to allow the cage inhabitants to completely submerge themselves. A branch or two will allow the opportunity for climbing.

FEEDING: Water skinks will feed upon moths, beetles, caterpillars, mealworms and other insects, as well as spiders, and commercial canned pet foods (stay clear of fish-types) in captivity. Adults should be fed every two to three days and juveniles every day.

CAPTIVE BEHAVIOUR: Specimens rapidly become used to human presence and will remain active when disturbed. In time they become amenable to handling.

BREEDING: Water skinks mate in the spring, following winter cooling, and give birth to two to seven young in December or January. The young are about 5cm long at birth , most of which is tail, and feed on the same diet as adults. Young individuals should be housed separately from the adults.

PINK-TONGUED SKINK
Tiliqua gerrardii

pink-tongue

HABITS: The pink-tongued skink is semi-arboreal, climbing low shrubs and trees but also spending much time on the ground. It can sometimes be found active by day but is largely nocturnal. It feeds mainly on snails and slugs, and has a specialized large tooth at the back of each jaw to crush snail shells.

HOUSING: In general, pink-tongued skinks are best kept indoors, although in warmer climates they have been successfully kept outdoors. Their ability as climbers makes it essential that outdoor housing be completely enclosed. Indoor caging for adult pink-tongues should provide a warm area for basking, preferably heated from above, although lighting should be arranged so that the lizards cannot climb the fixtures and burn themselves.

A suitable cage for a pair of adult pink-tongued skinks should have a length of at least 50cm and preferably enough height for decoration to indulge the lizard's climbing habits.

FEEDING: Pink-tongued skinks feed readily on snails and slugs, but can generally be readily weaned onto a diet of commercial tinned cat food (non-fish variety). If snails and slugs are fed, care should be taken to ensure that the source is entirely free of snail baits and garden poisons. Snails collected from the vicinity of Oleander bushes and other poisonous plants have also been known to cause fatalities to pink-tongues.

CAPTIVE BEHAVIOUR: Although initially very shy, pink-tongues soon become used to human presence. Handling should be minimal. Pink-tongues spend much of the day under shelter or basking, and are most active after dark on warm nights. Spraying the cage occasionally with a fine mist simulates rain and causes the lizards to actively search for snails brought out by the 'rain'.

BREEDING: Pink-tongued skinks breed readily in captivity, usually producing from 5-30 small live young in early summer (December-January), following winter cooling. The young are about 5-7cm total length at birth, about half of which is tail, and are much more boldly marked/coloured than adults. Young pink-tongues must be housed separately from adults and will often fight between themselves, especially over food.

BLOTCHED BLUETONGUE SKINK ***Tiliqua nigrolutea***
EASTERN BLUETONGUE SKINK ***T. scincoides***
(and other bluetongues)

blotched bluetongue

eastern bluetongue

HABITS: Bluetongues are completely ground-dwelling, sheltering at night and during the heat of the day in leaf litter, dense grass or beneath solid objects. During the early morning and mid-afternoon, they alternately bask in sunny patches, flattening out the body to gain maximum exposure, and forage for a wide range of foods, both plant and animal.

HOUSING: If the species occurs naturally in the area, it may be kept in a large outdoor enclosure with smooth walls extending at least 50 cm below ground and 100 cm above ground. Such an enclosure should be situated in a position that will receive more than 8 hours of sun a day, but should also provide some shaded and dry areas. During winter, a deep layer of dry grass and leaves should be provided in a dry place for the lizard to hibernate in. An indoor enclosure for one to two adults must be at least 75cm long and provide a warm area for basking. Shelter in the form of a low flat box, leaf litter or loose grass should also be provided.

FEEDING: In captivity these skinks will feed readily on a wide variety of foods, including soft fruits and vegetables, raw meat and commercial tinned pet foods (non-fish varieties). The diet should be varied occasionally. Snails and slugs will also be eaten though these should only be collected from areas without snail baits or poisonous plants. A once weekly feeding schedule is suitable.

CAPTIVE BEHAVIOUR: Although often timid and liable to bite when initially collected, these large skinks will rapidly become settled and are amenable to periodic handling, although they may hiss when approached.

BREEDING: Bluetongues mate in the spring following winter cooling and give birth to young in summer or early autumn. Blotched bluetongues produce two to ten young, while common bluetongues give birth to 5-18 young. New-born bluetongues are about 7cm long and readily feed on the same diet as adults. They should be housed separately from the adults.

SHINGLEBACK SKINK
Tiliqua* (or *Trachydosaurus*) *rugosa

shingleback

HABITS: Shinglebacks are completely ground-dwelling, sheltering at night and during the heat of the day in leaf litter, dense grass or beneath solid objects. During the early morning and mid-afternoon, they alternately bask in sunny patches and forage. Shinglebacks are slow moving and are often seen basking on roads during the day. They will eat anything slow enough to catch, as well as a large proportion of plant material.

HOUSING: In areas where they occur naturally, shinglebacks may be kept in a large outdoor enclosure such as fenced off gardens. Enclosure fences should extend at least 50 cm below ground and 80 cm above ground. Such an enclosure should be situated in a position that will receive more than 8 hours of sun a day, but should also provide some shaded and dry areas. During winter, a deep layer of dry grass and leaves should be provided in a dry place for the lizard to hibernate in.

In areas of high humidity, such as on the east coast, shinglebacks suffer respiratory problems, and must therefore only be kept indoors. An indoor cage for one to two adults must be at least 100cm long and provide a warm area for basking, preferably heated by a Gro-Lite incandescent globe or equivalent. Shelter in the form of a low flat box, leaf litter or loose grass should also be provided. Unlike caging for most other reptiles, no water dish should be left in the cage. Instead, the lizards should be allowed the opportunity to drink for a short time every one to two weeks, and the water bowl then removed.

FEEDING: In captivity shinglebacks feed readily on a wide variety of foods, including soft fruits and vegetables, raw meat and commercial tinned pet foods (non-fish varieties), but concentrating on vegetables. Occasional supplementation of the diet with calcium powder and vitamin drops is suggested. Shinglebacks will also readily eat snails and slugs, though these should only be collected from areas without snail baits or poisonous plants. A once weekly feeding schedule is suitable.

CAPTIVE BEHAVIOUR: Although often timid and liable to bite when initially collected, shinglebacks rapidly become settled and allow periodic handling.

BREEDING: Shinglebacks give birth to one to four (usually two or three) young in summer or early autumn. Mating occurs in spring. New-born young are about 9 cm long and readily feed on the same diet as adults.

CHILDREN'S PYTHON *Liasis childreni*
(and other small python species)

Children's python

HABITS: Children's pythons are mainly nocturnal and are not particularly fond of climbing trees, though they do climb amongst rock formations, often sheltering in crevices or in burrows beneath. Instead of foraging, Children's pythons often wait in in ambush, ready to strike at any passing lizard or mammal of appropriate size.

HOUSING: An indoor terrarium at least 60cm long is needed to house one or two adult specimens. Cage decorations should include adequate hiding places such as appropriately sized and shaped pieces of bark upon the cage floor, situated in a warm area of the cage.

To stimulate nocturnal activity of specimens and to improve your ability to watch them, a red or blue light globe may be installed to light and warm the cage by night. The cage may be warmed by day using the same globe or by installing a second light fitting for a normal white light (or ultraviolet) globe of appropriate wattage to be turned on during daylight hours as needed. A gradual cline of temperatures from one end of the cage to the other is preferred and can be promoted by placing the heat source towards one end of the cage. Temperature readings from the middle of the cage should be maintained between 24 and 28°C.

FEEDING: Most adult specimens will readily accept mice and lizards though some specimens of inland species may insist upon an exclusive menu of lizards. Juveniles often require skinks initially. Meals should be offered about once a week or fortnight and as this species is prone to obesity, meals should be small if offered weekly.

CAPTIVE BEHAVIOUR: Newly hatched or captured specimens will quickly settle in to become extremely hardy in captivity. They are generally amenable to handling.

BREEDING: These species are amongst the easiest of Australian snakes to breed. Adults may be kept together year round and expected to produce an annual clutch of approximately a dozen eggs, usually in the summer following spring mating. Sex of juveniles and adults may be determined by probing (preferably by someone experienced in probing small snakes).

CARPET PYTHON/DIAMOND PYTHON *Morelia spilota*

carpet python

diamond python

HABITS: Carpet pythons and diamond pythons are highly arboreal, and though fairly nocturnal, are quite often discovered while active by day, either sun basking or, in the springtime, when males search widely for females to mate with. Nocturnal foraging for birds and mammals is done in trees and shrubs, as well as at ground level, though juveniles are believed to feed largely on lizards. Feeding also takes place by ambush. Inactive periods are spent within retreats such as hollow limbs or rock crevices or coiled on a branch.

HOUSING: To house one or two adult specimens, an indoor terrarium of minimal dimensions 50cm x 100cm x 100cm should be provided. It is recommended that the cage be oriented such that the height is not the smaller of these dimensions. Cage decorations should include branches for climbing, and hiding places such as cardboard, wooden or plastic boxes with a small entryway cut out, or hollow logs, etc, situated in a warm area of the cage.

To stimulate nocturnal activity of specimens and to improve your ability to watch them, a red or blue light globe may be installed to light and warm the cage by night. The cage may be warmed by day using the same globe or by installing a second light fitting for a normal white light (or ultraviolet) globe of appropriate wattage to be turned on during daylight hours as needed. A gradual cline of temperatures from one end of the cage to the other is preferred and can be promoted by placing the heat source towards one end of the cage. The temperature at the middle of the cage should be maintained between 24 and 28°C.

FEEDING: Warm-blooded prey such as rodents, chickens, etc, will be readily accepted by hungry specimens. Meals should be offered about once a week or fortnight. Juveniles may require skinks initially. This species is usually over-fed in captivity and meals should be small if fed weekly.

CAPTIVE BEHAVIOUR: Newly captured specimens may be nervous, in which case they will not hesitate to bite savagely, but most specimens will quickly settle in to become extremely hardy in captivity and are generally amenable to handling.

BREEDING: This species is egg laying, producing 12-50 eggs in the summer. Mating occurs in the spring, following winter cooling. Sex of juveniles may be determined by probing (preferably by someone experienced in probing small snakes), but adults shouldn't be probed as this will very readily cause injury in females. Males differ from females in having a much more prominent claw on the end of their paired cloacal spurs (see diagram in breeding section of book).

BLACK-HEADED PYTHON *Aspidites melanocephalus*
WATER PYTHON *Liasis fuscus*
OLIVE PYTHON *L. olivaceus*
SCRUB PYTHON *Morelia amethistinus*
(and other large python species)

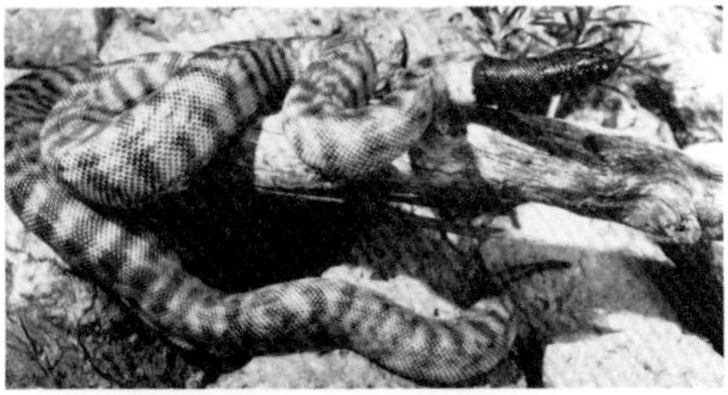

black-headed python

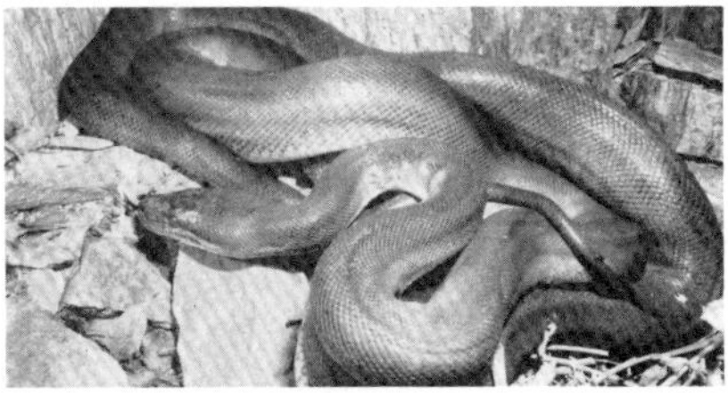

olive python

water python

scrub python

HABITS: Scrub pythons are largely arboreal in the wild, olive pythons and water pythons are partially arboreal, (olive pythons also frequenting rock outcrops and caves), while black-headed pythons are mainly ground dwelling. All four species are mainly nocturnal, though at times, all to a greater or lesser extent may be discovered sun basking or otherwise active by day. All are most frequently encountered when crossing rural roads at night. Water pythons and scrub pythons feed mainly upon warm-blooded prey, while black-headed pythons and olive pythons are less choosy and will eat practically any reptile, mammal or bird which can be killed and swallowed. All species kill their food by constriction and are capable of swallowing prey items several times their own girth.

HOUSING: Except during periods of introduction for the purpose of mating, all large specimens are best kept in separate cages to avoid problems during feeding (which might include cannibalism). To house one, or occasionally two specimens, an indoor terrarium should be at least half the length of the longest specimen within; the minimum width should be half the measurement of the cage length. For all species except the black-headed python, the enclosure should be tall enough to accommodate the inclusion of sturdy branches for climbing. Apart from branches for climbing, cage furnishings should include hiding places, such as cardboard, wooden or plastic boxes with a small entryway cut out, or hollow logs, etc, situated in a warm area of the cage. To stimulate nocturnal activity of specimens and to improve your ability to watch them, a red or blue light globe may be installed to light and warm the cage by night. The cage may be warmed by day using the same globe or by installing a second light fitting for a normal white light (or ultraviolet) globe of appropriate wattage to be turned on during daylight hours as needed. A gradual cline of temperatures from one end of the cage to the other is preferred and can be promoted by placing the heat source towards one end of the cage. The temperature in the middle of the cage should be maintained between 24 and 28°C.

FEEDING: Warm-blooded prey such as rodents, chickens, etc, will be readily accepted by hungry specimens. Meals should be offered about once a week or fortnight. Juveniles may require skinks or small birds (eg finches) initially.

CAPTIVE BEHAVIOUR: Newly captured specimens may be nervous, in which case they will not hesitate to bite savagely, but will quickly settle in to become extremely hardy in captivity. They are generally amenable to handling.

BREEDING: All pythons lay eggs, usually in summer following spring mating, though these events may occur somewhat earlier. Water pythons breed very readily, though the other three are less often bred. Sex of juveniles and adults may be determined by probing (preferably by someone experienced in probing small snakes).

BROWN TREE SNAKE *Boiga irregularis*

brown tree snake

NOTE: This is a venomous species, although its bite is not considered to be dangerous to humans.

HABITS: The brown tree snake is a nocturnal arboreal species. It forages at night in trees, amongst rock outcrops and on the ground for birds, small mammals and lizards. When inactive, this species retreats into hollow limbs and rock crevices.

HOUSING: To house one or two adult brown tree snakes you will need an indoor terrarium of at least 90cm length that is tall enough to include branches for climbing. Cage decoration should include suitable arrangements of branches, and hiding places (such as small cardboard boxes with a small entryway cut out, or appropriately sized and shaped pieces of bark upon the cage floor) situated in a warm area of the cage. To stimulate nocturnal activity of specimens and to improve your ability to watch them, a red or blue light globe may be installed to light and warm the cage by night. The cage may be warmed by day using the same globe or by installing a second light fitting for a normal white light (or ultraviolet) globe of appropriate wattage to be turned on during daylight hours as needed. A gradual cline of temperatures from one end of the cage to the other is preferred and can be promoted by placing the heat source towards one end of the cage. The temperature in the middle of the cage should be maintained between 24 and 28°C.

FEEDING: Brown tree snakes adapt readily to a diet of small rodents or small birds. Meals should be offered about once a week.

CAPTIVE BEHAVIOUR: Newly captured specimens are nervous, but will quickly settle in to become extremely hardy in captivity, although they are generally not amenable to handling.

BREEDING: This species is egg laying, producing three to nine eggs in the late spring or summer following winter cooling. Sex may be determined by probing.

COMMON TREE SNAKE *Dendrelaphis punctulatus*

common tree snake

HABITS: The green tree snake is a very active diurnal arboreal species. It forages in trees and shrubs, as well as on the ground during the day for its major food source, frogs. It also is known to eat skinks and fish. When inactive, it shelters in hollow limbs, rock crevices or beneath rock exfoliations.

HOUSING: A terrarium at least 90cm long is required for 1-2 adult green tree snakes and must be tall enough to include branches for climbing. Cage decoration should include suitable arrangements of branches, and hiding places such as small cardboard, wooden or plastic boxes with a small entryway cut out, or appropriately sized and shaped pieces of bark upon the cage floor, situated in a warm area of the cage.

The cage should be warmed by day, preferably by a radiant heat source (ie. a reflector globe) to allow basking. This should be placed at one end of the cage to provide a thermal gradient. Temperature readings from the middle of the cage should be maintained at 24-28°C.

FEEDING: Juvenile green tree snakes may need to be fed on frogs, large tadpoles or skinks, but adults can usually be weaned onto a diet of young mice or small, whole fish (without spines). Meals should be offered about once a week.

CAPTIVE BEHAVIOUR: Newly captured specimens are highly nervous, but will quickly settle in to become extremely hardy in captivity. Green tree snakes are amenable to occasional handling.

BREEDING: This species is egg laying, producing 7-11 eggs in the spring. Sex may be determined by probing.

DEATH ADDER ***Acanthophis antarcticus***

death adder

CAUTION: The death adder is dangerously venomous and should only be handled with the appropriate tools and with great caution. There is no excuse for any reptile keeper to be bitten. However, if a bite does occur, regardless of how small the specimen or how minor the bite might appear to be, the injury must be immediately bandaged (as diagramed in this book, p 37), and medical assistance sought. Dangerously venomous species should only be kept by persons of at least 18 years of age, and where adequate prior experience has been gained in the keeping and handling of harmless species.

HABITS: The death adder moves about most frequently by night or during warm, overcast conditions and is most often encountered when it crosses roads at night. During both day and night, the majority of time is spent concealed amongst or beneath leaf litter awaiting the approach of potential prey animals such as lizards, birds and small mammals. These are lured to within striking distance by writhing the specialized tail tip in a manner that mimics a small worm or insect.

HOUSING: As a dangerously venomous species, the death adder should only be kept in accordance with the minimum caging standards for dangerous snakes which are described and illustrated in this book. One or two adult snakes can be kept in an indoor terrarium with a minimum length of 60cm. Juvenile specimens (less than 25cm length) may be kept in smaller facilities but minimum safety requirements should be met, and gaps of greater than 1mm are not advised. Cage decoration should include a leaf-litter substrate.

To stimulate nocturnal activity of specimens and to improve your ability to watch them, a red or blue light globe may be installed to light and warm the cage by night. The cage may be warmed by day using the same globe or by installing a second light fitting for a normal white light (or ultraviolet) globe of appropriate wattage to be turned on during daylight hours as needed. A gradual cline of temperatures from one end of the cage to the other is preferred and can be promoted by placing the heat source towards one end of the cage. The temperature in the middle of the cage should be intermediate, and maintained between 24 and 28°C.

FEEDING: Adults will generally accept small rats and mice in captivity, though some individuals require large skinks or birds. Juveniles may need to be fed initially on skinks, but can eventually be weaned onto a diet of young mice. All death adders must either be separated from one another during feeding or watched carefully to avoid accidental cannibalism (when more than one specimen attempts to swallow the same food item). Meals should be offered about once a week. Captive death adders typically become obese in a short time, and keepers should consider using smaller food items or feeding less often if this problem arises.

CAPTIVE BEHAVIOUR: Death adders adjust extremely well to captivity and if handled only in a safe manner with appropriate tools, such as snake hooks, make an ideal choice for many keepers. However, it is important not to underestimate the ability of death adders to strike rapidly and without warning in any direction, including directly upwards if hungry and given the opportunity. Hands must never get within striking distance.

BREEDING: Sex may be readily determined by tail shape, males having a very much thicker tail base than females. Death adders mate in spring and usually produce 3-19 young between February and April. Young are approximately 13cm long at birth.

COPPERHEAD *Austrelaps superbus*
TIGER SNAKE *Notechis scutatus*

copperhead

tiger snake

CAUTION: Copperheads and tiger snakes are dangerously venomous and should only be handled with the appropriate tools and with great caution. There is no excuse for any reptile keeper to be bitten. However, if a bite does occur, regardless of how small the specimen or how minor the bite might appear to be, the injury must be immediately bandaged (as diagramed in this book, p 37), and medical assistance sought. Dangerously venomous species should only be kept by persons of at least 18 years of age, and where adequate prior experience has been gained in the keeping and handling of harmless species.

HABITS: Large populations of both species often live in proximity to one another in and around swamps and marshes. Both are active mainly by day. Copperheads often spend nights in grassy tussocks while tiger snakes more often seek shelter within fallen hollow logs rock piles or in burrows. During warm months both species can be seen in the morning and periodically throughout the day, basking in the sun. When necessary, the entire body can be flattened to gain maximum exposure to the sun. Basking behavior is alternated with episodes of foraging, during which bushes, grass tussocks are explored for prey. Copperheads feed mainly on frogs and skinks, while tiger snakes, in addition to these food types, also feed extensively on small mammals and nestling birds.

HOUSING: As dangerously venomous species, copperheads and tiger snakes should only be be kept in accordance with the minimum caging standards for dangerous snakes which are described and illustrated in this book. Two adult snakes can be kept in an indoor terrarium with a minimum length of 90cm. Single, average sized specimens may be adequately accommodated by a floor area measuring 60cm x 60cm. Juvenile specimens (less than 40cm length) may be kept in smaller facilities but minimum safety requirements should be met, and gaps of greater than 1mm are not advised. Cage decoration should include suitable hiding places such as cardboard, wooden or plastic boxes with a small entryway cut out, situated in a warm area of the cage.

The cage must be heated by day, and this is best accomplished by including a radiant heat source (ie. a reflector globe) towards one end of the cage to allow basking. For copperheads, the temperature readings from the middle of the cage should be cooler than recommended for other species, and maintained between 22 and 26°C. Copperheads are adapted to surviving in relatively cool regions, and in captivity too much warmth can result in maladjustment and eventual illness. During the night, all heating devices should be turned off and the cage temperature allowed to drop considerably. Tiger snakes should be kept warmer, temperature readings from the middle of the cage should be intermediate, and maintained between 24 and 28°C. Night-time temperatures should not be allowed to drop significantly.

FEEDING: Copperheads and young tiger snakes may need to be fed initially on frogs or possibly skinks, but adults of both species can readily be weaned onto a diet of young mice if these are not initially accepted. Consult the feeding section of this book for techniques for weaning snakes onto rodent diets. Meals should be offered about once a week.

CAPTIVE BEHAVIOUR: Copperheads adjust readily to captivity and are likely to become fairly placid in a short time. Freshly collected tiger snakes are apt to be very defensive and if given the opportunity, will bite. In time however, the defensive exhibitions towards the keeper wane and they adapt well to captivity. Of course specimens must be handled only with appropriate tools, such as snake hooks.

BREEDING: Sexing adult snakes can be done by tail shape or by probing (by a competent snake handler) as described in the breeding section of this book. Mating takes place mainly in the autumn as well as in the spring for both species, and pairs should be introduced at that time if breeding is desired. The young (averaging 20 in copperheads and 30 in tiger snakes) are usually born during the summer (tiger snakes) or autumn (copperheads) and are very small (14-20 cm length). The young should be handled with a small snake hook and immediately moved into a smaller escape-proof cage until disposed of, or in the event that any are to be kept, until they exceed 50 cm in length, at which time the minimum cage dimensions for adults must be met. The young must be housed separately from adults, or they will be eaten.

YELLOW-FACED WHIP SNAKE *Demansia psammophis*

yellow-faced whip snake

CAUTION: The yellow-faced whip snake is superficially similar to the common brown snake *Pseudonaja textilis*, which is dangerously venomous. Great care should therefore be taken to ensure correct identification. Though not regarded as particularly dangerous, the bite of a yellow-faced whip snake can be very painful, and severe cases may require medical attention.

HABITS: The yellow-faced whip snake is a very active, terrestrial species, foraging widely during the day for its major food source - skinks. When inactive, this species can often be found beneath rock exfoliations, bits of timber or rubbish (disposed sheets of tin, boards), etc.

HOUSING: One or two adult yellow-faced whip snakes require an enclosure at least 70cm long. Cage decoration should include suitable hiding places such as small boxes with an entryway cut out, or appropriately sized and shaped pieces of bark. Hiding places should be situated in a warm area of the cage.

Cages should be warmed by day, preferably by a radiant heat source (ie. a reflector globe) to allow basking. This should be placed at one end of the cage to provide a thermal gradient. The temperature in the middle of the cage should be maintained at 24-28°C.

FEEDING: Yellow-faced whip snakes feed mainly on skinks, but may also accept frogs. It may also be possible to wean specimens onto a diet of young mice. Meals should be offered about once a week.

CAPTIVE BEHAVIOUR: Newly captured specimens are highly nervous but will quickly settle in to become extremely hardy in captivity.

BREEDING: This species is egg laying, producing 3-9 eggs in summer, following winter cooling. Sex may be determined by tail shape or probing.

DE VIS' BANDED SNAKE *Denisonia devisii*
CURL SNAKE *Suta suta*

De Vis' banded snake

curl snake

CAUTION: Care must be taken that the De Vis' banded snake is not confused with the common death adder *Acanthophis antarcticus*, a dangerously venomous species which it superficially resembles. Both the De Vis' banded snake and the curl snake are pugnacious when first captured and will not hesitate to bite, and although not regarded as dangerous, bites can be very painful, and in the case of large specimens, should be medically treated.

HABITS: De Vis' banded snakes and curl snakes are ground dwelling species and most active at dusk or after dark on warm evenings, though curl snakes are also active on cool breezy nights when few reptiles are active. In the wild, De Vis' banded snakes feed mainly on frogs, while curl snakes concentrate on lizards and the occasional frog or mouse. By day both species can be found sheltering beneath logs, railway sleepers and other debris.

HOUSING: An adult pair of either species may be adequately kept in an indoor terrarium at least 50cm long. Cage decoration should include suitable hiding places (such as appropriately sized and shaped pieces of bark, etc) for shelter situated in a warm area of the cage.

To stimulate nocturnal activity of specimens and to improve your ability to watch them, a red or blue light globe may be installed to light and warm the cage by night. The cage may be warmed by day using the same globe or by installing a second light fitting for a normal white light (or ultraviolet) globe of appropriate wattage to be turned on during daylight hours as needed. A gradual cline of temperatures from one end of the cage to the other is preferred and can be promoted by placing the heat source towards one end of the cage. The temperature in the middle of the cage should be maintained between 24 and 28°C.

FEEDING: Meals should be offered about once a week. De Vis' banded snakes readily take frogs and may be weaned onto a diet of young mice. Curl snakes readily accept lizards, and can normally be weaned onto a diet of young mice.

CAPTIVE BEHAVIOUR: Both species will adjust readily to captivity but should not be regularly handled other than with a snake hook. When disturbed, they will frequently thrash about wildly and readily bite.

BREEDING: De Vis' banded snakes produce three to nine young in late summer. Curl snakes produce two to seven young, also in summer. Sex may be determined by tail shape or by probing.

SWAMP SNAKE ***Hemiaspis signata***

swamp snake

CAUTION: Swamp snakes are superficially similar to juvenile common brown snakes *Pseudonaja textilis*, which are dangerously venomous. Great care should therefore be taken to ensure correct identification. Though not regarded as dangerous, the bite of a swamp snake may be painful and cause a significant local reaction, and severe cases may require medical attention.

HABITS: The swamp snake is a diurnal, terrestrial species often found beneath discarded rubbish, sheets of building materials, etc, around rural properties.

HOUSING: To house a pair of adult swamp snakes you will need an indoor terrarium at least 50cm long. Cage decoration should include suitable hiding places such as appropriately sized and shaped pieces of bark upon the cage floor, situated in a warm area of the cage.

You will need to warm the cage by day, and this is best accomplished by including a radiant heat source (ie. a reflector globe) to allow basking. A gradual cline of temperatures from one end of the cage to the other is preferred and can be promoted by placing the prime heat source towards one end of the cage. The temperature in the middle of the cage should be intermediate and maintained between 24 and 28°C.

FEEDING: Swamp snakes generally need to be fed frogs or skinks, but can occasionally be weaned onto a diet of young mice. Meals should be offered about once a week.

CAPTIVE BEHAVIOUR: Swamp snakes are very hardy in captivity and are amenable to occasional handling, though if they do bite the result may be painful.

BREEDING: This species is live bearing, producing 6-16 young in late summer or autumn. Sex may be determined by tail shape or probing.

PALE-HEADED SNAKE *Hoplocephalus bitorquatus*

pale-headed snake

CAUTION: The pale-headed snake is regarded as potentially dangerous and should be housed only in escape-proof facilities. This species is very agile and will bite readily. All bites should be treated as serious, and medical attention sought.

HABITS: The pale-headed snake is a nocturnal, arboreal species. When inactive, it shelters beneath loose, exfoliating bark on trees. At night it prowls in trees and on the ground for frogs and lizards.

HOUSING: A pair of adult adult pale-headed snakes should be housed in an indoor terrarium at least 50cm long, and tall enough to include branches for climbing. Cage decoration should include suitable hiding places (such as small cardboard boxes with a small entryway cut out) situated in a warm area of the cage.

To stimulate nocturnal activity of specimens and to improve your ability to watch them, a red or blue light globe may be installed to light and warm the cage by night. The cage may be warmed by day using the same globe or by installing a second light fitting for a normal white light (or ultraviolet) globe of appropriate wattage to be turned on during daylight hours as needed. A gradual cline of temperatures from one end of the cage to the other is preferred and can be promoted by placing the heat source towards one end of the cage. The temperature in the middle of the cage should be maintained between 24 and 28°C.

FEEDING: Pale-headed snakes may need to be fed initially on frogs or skinks, but can usually be weaned onto a diet of young mice. Meals should be offered about once a week.

CAPTIVE BEHAVIOUR: Pale-headed snakes adjust readily to captivity, but because they are small, agile and inclined to bite at any opportunity, specimens should be handled with particular caution, and only with appropriate tools, such as snake hooks.

BREEDING: Two to eleven young are produced, usually in February. Sex may be determined by probing by a highly experienced handler.

KING BROWN SNAKE (mulga snake) ***Pseudechis australis***
COLLETT'S SNAKE ***P. collettii***
SPOTTED OR BLUE-BELLIED BLACK SNAKE ***P. guttatus***

king brown snake

Collett's snake

spotted black snake

CAUTION: King brown snakes, Collett's snakes and spotted black snakes are all dangerously venomous and should only be handled with the appropriate tools and with great caution. The venoms of these three species are particularly damaging to tissue and severe complications typically follow bites. There is no excuse for any reptile keeper to be bitten. However, if a bite does occur, regardless of how small the specimen or how minor the bite might appear to be, the injury must be immediately bandaged (as diagramed in this book, p 37), and medical assistance sought. Dangerously venomous species should only be kept by persons of at least 18 years of age, and where adequate prior experience has been gained in the keeping and handling of harmless species.

Most importantly, keepers must be aware that all three species become very dangerous at feeding time (or whenever keeper activity might be confused with the introduction of food). At such times specimens become highly excited, thrashing about the cage and biting any object in their path. If movement is perceived through an opened cage door, a specimen may lunge out of the cage unexpectedly in an attempt to bite the source of movement - the keeper.

HABITS: King brown snakes and Collett's snakes are primarily nocturnal and are usually collected when crossing roads on warm nights. When inactive, they shelter in animal burrows. Spotted black snakes are active by night and day, and are often collected under sheets of metal, railway sleepers, or other debris, or when basking. Natural prey items include frogs, reptiles (including snakes) and warm-blooded prey small enough to swallow.

HOUSING: As dangerously venomous species, they should only be be kept in accordance with the minimum caging standards for dangerous snakes which are described and illustrated in this book. Cage size will depend upon the size of the specimens being kept. Specimens up to 1.8 metres will do well in an indoor terrarium with a length of 100cm. It is recommended that each specimen be housed separately (except during intervals designated for mating), as great difficulty can occur during feeding time or when the snakes are hungry. These species are unusual in that they may be killed if envenomated by another member of their species; death occurring one or two weeks after the bite is inflicted.

Juvenile specimens (less than 40cm length) are best kept in smaller facilities than adults, but minimum safety requirements should be met, and gaps of greater than 1mm are not advised. Cage decoration should include suitable hiding places situated in a warm area of the cage.

To stimulate nocturnal activity of specimens and to improve your ability to watch them, a red or blue light globe may be installed to light and warm the cage by night. The cage may be warmed by day using the same globe or by installing a second light fitting for a normal white globe to be turned on during daylight hours as needed. A cline of temperatures from one end of the cage to the other is preferred and can be promoted by placing the heat source towards one end of the cage. The temperature in the middle of the cage should be intermediate and maintained between 24-28°C.

FEEDING: Adults will generally accept either rats or mice in captivity. Juveniles may need to be fed initially on frogs or skinks, but can readily be weaned onto a diet of young mice. All specimens must be separated from one another during feeding or watched carefully to avoid accidental cannibalism when more than one specimen attempts to swallow the same food item. Meals should be offered about once a week.

CAPTIVE BEHAVIOUR: These species adjust very well to captivity and will become largely diurnal in habits if a radiant heat source is provided to allow basking. Handle only with appropriate tools, such as snake hooks and do not underestimate the ability of these dangerous snakes to strike unexpectedly (especially in a sideways, wide sweeping manner) when given the opportunity.

BREEDING: All three are egg-laying species that are readily bred in captivity. Mating takes place in the spring and pairs should be introduced at that time if breeding is desired. In the case of king brown snakes, the eggs are laid only a short time after mating (about 45 days), while a period of about 60 days is typical for Collett's snakes and spotted black snakes. The eggs for all three species take about two months to hatch, depending upon incubation temperature. Sex may be determined by tail shape or by probing by an experienced handler.

RED-BELLIED BLACK SNAKE
Pseudechis porphyriacus

red-bellied black snake

CAUTION: The red-bellied black snake is dangerously venomous and should only be handled with the appropriate tools and with great caution. Although the venom is less toxic than that of other dangerously venomous elapids, severe reactions have been seen in several cases. There is no excuse for any reptile keeper to be bitten. However, if a bite does occur, regardless of how small the specimen or how minor the bite might appear to be, the injury must be immediately bandaged (as diagramed in this book, p 37), and medical assistance sought. Dangerously venomous species should only be kept by persons of at least 18 years of age, and where adequate prior experience has been gained in the keeping and handling of harmless species.

HABITS: Red-bellied black snakes are primarily diurnal and active during the warmer months. During the morning and periodically throughout the day they can be found basking in the sun with nearly the entire body flattened out in an effort to gain maximum exposure to the sun. Basking behavior is alternated with episodes of foraging, during which bushes, grass tussocks, burrows and ground crevices are explored for frogs, lizards and small mammals.

HOUSING: As a dangerously venomous species, the red-bellied black snake should only be be kept in accordance with the minimum caging standards for dangerous snakes which are described and illustrated in this book. Two adult snakes can be kept in an indoor terrarium with a minimum length of 90cm. Single, average sized specimens may be adequately accommodated by a floor area measuring 60cm x 60cm. Juvenile specimens (less than 40cm length) may be kept in smaller facilities but minimum safety requirements should be met, and gaps of greater than 1mm are not advised. Cage decoration should include suitable hiding places such as cardboard, wooden or plastic boxes with a small entryway cut out, situated in a warm area of the cage.

The cage must be heated by day, and this is best accomplished by including a radiant heat source (ie. a reflector globe) at one end of the cage to allow basking. The temperature in the middle of the cage should be intermediate, and maintained between 24 and 28°C.

FEEDING: Adults will generally accept mice in captivity. Juveniles may need to be fed initially on frogs or skinks, but can readily be weaned onto a diet of young mice. Meals should be offered about once a week.

CAPTIVE BEHAVIOUR: Red-bellied black snakes will adjust readily to captivity and are likely to become fairly placid in a short time. Once 'settled in', they are amongst the most suitable of captive snakes. However, even such 'tame' individuals still must be handled only with appropriate tools, such as snake hooks.

BREEDING: This species is readily bred in captivity. Mating takes place in the spring, and pairs should be introduced at that time if breeding is desired. The young (numbering 8-40) are born during the summer or autumn. The young should be handled with a small snake hook and immediately moved into a smaller escape-proof cage until disposed of, or in the event that any are to be kept, until they exceed 50cm in length, at which time the minimum cage dimensions for adults should be met. Sex may be determined by tail shape or by probing by an experienced handler.

WESTERN BROWN SNAKE *Pseudonaja nuchalis*

western brown snake

CAUTION: Western brown snakes are dangerously venomous and should only be handled with the appropriate tools and with great caution. There is no excuse for any reptile keeper to be bitten. However, if a bite does occur, regardless of how small the specimen or how minor the bite might appear to be, the injury must be immediately bandaged (as diagramed in this book, p 37), and medical assistance sought. The symptoms of western brown snake bites may be slow to appear, but their onset quite dramatic. Dangerously venomous species should only be kept by persons of at least 18 years of age, and where adequate prior experience has been gained in the keeping and handling of harmless species.

HABITS: Frequently encountered when crossing roads during the day or at dusk, western brown snakes are a common sight for rural residents as they frequently inhabit properties. When inactive, shelter is sought within animal burrows or under timber. Natural prey items include skinks and mice.

HOUSING: As a dangerously venomous species, the western brown snake should only be be kept in accordance with the minimum caging standards for dangerous snakes which are described and illustrated in this book. One to two adults can be kept in a terrarium with a minimum length of 90cm. Juvenile specimens (less than 40cm length) may be kept in smaller facilities but minimum safety requirements should be met, and gaps of greater than 1mm are not advised. Cage decoration should include suitable hiding places situated in a warm area of the cage. The cage must be heated by day, and this is best accomplished by including a radiant heat source (reflector globe) at one end of the cage to allow basking. The temperature in the middle of the cage should be intermediate and maintained between 26 and 30°C.

FEEDING: Adults will generally accept either rats or mice in captivity. Juveniles may need to be fed initially on skinks, but can eventually be weaned onto a diet of young mice. Meals should be offered about once a week.

CAPTIVE BEHAVIOUR: Western brown snakes are fairly shy, and may require a good deal of quiet before settling into captivity. This species will readily bite if disturbed, and must only be handled with appropriate tools, such as snake hooks.

BREEDING: Sex may be determined by probing by an experienced handler. This is an egg-laying species, laying between 24-38 eggs per clutch.

BROAD-SHELLED TORTOISE *Chelodina expansa*

broad-shelled tortoise

HABITS: The broad-shelled tortoise is rather secretive, often laying in shallow water with its shell submerged in sand. Here it waits for small fish or other small animals to swim past and snaps them up with a sudden lunge of the snake-like head and neck.

HOUSING REQUIREMENTS: For adult specimens, outdoor accommodation is necessary. It is essential that plenty of both sunshine and shade are available in the water and on the land provided for the tortoises, and that no sharp or rough edges exist on the surface of the pool if it is constructed from concrete.

It is essential that broad-shelled tortoises are not overcrowded. A water area of 150cm x 200cm with a depth of 50cm is required for one or two adults, with another 150cm x150cm surface area added for each additional specimen intended to share the pool. An equal ground area should be made available. The wall bordering the enclosure should be sunk at least 50cm below the ground to prevent escape by digging specimens. The walls must be smooth and at least a metre high. A pile of loose vegetation cuttings should be made available during the cooler months, beneath which the tortoises will bury themselves while hibernating.

Juveniles should be kept in indoor aquaria or aqua-terraria with the water temperature maintained at 22-26°C. A platform or equivalent must be made available for the tortoises, upon which they may dry themselves and bask under a radiant heat source. Unless the tank receives adequate sunlight through a nearby open window, an artificial source of ultraviolet light will be required, or supplementation of the diet with vitamin D and calcium will be necessary.

FEEDING: Broad-shelled tortoises will readily feed upon raw meat, insects (and other invertebrates), small mice, small whole fish, etc. If raw meat or less than whole prey animals are to comprise a major portion of any tortoise's diet, it should be supplemented once a week or so with calcium and vitamins. Live fish can be introduced to a tortoise pond as a supplemental food source. Uneaten food should not be left to foul the water.

CAPTIVE BEHAVIOUR: Although timid when first collected, this species adjusts well to captivity. Specimens will eventually take food directly from the hand.

BREEDING: Adult broad-shelled tortoises can usually be sexed on the basis of the shape of the plastron. In males the plastron is gently sculptured in a concave manner, lending stability during mating. As many as 18 eggs may be laid (usually buried near the tortoise's pond). These may take over a year to hatch. Both nesting and emergence of the young seem to be triggered by recent rain, which makes digging much easier. The hatchlings will need to moved to an isolated facility immediately or they may fall prey to the adults.

LONG-NECKED TORTOISE *Chelodina longicollis*

long-necked tortoise

HABITS: The long-necked tortoise may often be observed foraging close to shore or basking on the banks of freshwater sources. Wild specimens feed mainly upon animal matter, rarely eating any plant matter. Overland journeys for this species during summer months are common. Newly captured specimens may emit a strong defensive odour.

HOUSING REQUIREMENTS: For adult specimens, outdoor accommodation is preferable. It is essential that sunshine and shade are available in the water and on the land provided for the tortoises, and that no rough edges exist on the bottom and sides of the pool if it is constructed from concrete.

A water area of 100cm x 150cm with a depth of 40cm is required for one or two adults, with a further 100cm x100cm surface area added for each additional specimen intended to share the pool. An equal area of ground surface should be made available for each tortoise. The wall bordering the enclosure should be sunk at least 50cm below the ground, be smooth and at least a metre high. An inturned lip along the top of the perimeter wall is advised. Loose soil or, preferably, a pile of loose vegetation cuttings should be made available during the cooler months, beneath which the tortoises will bury themselves while hibernating.

Juveniles should be, and adults may be, kept in indoor aquaria or aqua-terraria with the water temperature being maintained between 23 and 26°C. A platform or equivalent must be made available for the tortoises, upon which they may bask under a radiant heat source. Unless the tank receives adequate sunlight through a nearby open window, an artificial source of ultraviolet light will be required, or supplementation of the diet with vitamin D and calcium will be necessary.

FEEDING: Long-necked tortoises will readily feed upon raw meat, insects (and other invertebrates), small mice, small whole fish, etc. If raw meat or less than whole prey animals are to comprise a major portion of any tortoise's diet, it should be supplemented once a week or so with a mixture of calcium and multiple vitamins. Live fish can be introduced to large tortoise ponds as a supplemental food source. They will elude the tortoises sufficiently to develop in numbers to a sizeable population. Tortoises normally take food only in the water and can only swallow it below the surface, but in time will learn to take food from your fingers before returning to the water to eat it. Uneaten food should not be left to foul the water.

CAPTIVE BEHAVIOUR: Long-necked tortoises are particularly suited to captivity and adjust almost spontaneously. They are also probably more resistant to diseases than any other tortoise.

BREEDING: There is no sure means of determining the sexes of this species. Usually, males will have a slightly concave plastron, allowing for stability when mating with females from above. Females tend to get larger than males. Nesting takes place in the spring and early summer, with clutches of up to 24 eggs being recorded. The eggs will typically be buried near the tortoise's pond and after an incubation period of approximately 3-4 months, the hatchlings will appear. They should be moved to an isolated facility immediately or they may fall prey to the adults.

MACQUARIE TORTOISE *Emydura macquarii*

Macquarie tortoise

HABITS: The Macquarie tortoise is a particularly active species. It may be observed foraging along the banks of rivers and amongst reeds, searching for small animals of any description, as well as some forms of plant matter. Overland journeys during summer months are not uncommon.

HOUSING REQUIREMENTS: For adult specimens, outdoor accommodation is preferable. It is essential that both sunshine and shade are available in the water and on the land provided for the tortoises, and that no sharp or rough edges exist on the surface of the pool if it is constructed from concrete. Also, a log should be suspended across the bottom and sides of the pond, or should emerge from it in such a way as to provide an accessible basking site above the water's surface. There should be enough basking space on these logs to accommodate all Macquarie tortoises in the enclosure.

A water area of 100cm x 150cm with a depth of 40cm is required for one or two adults, with another 100cm x100cm surface area added for each additional specimen intended to share the pool. An equal area of ground surface should be made available. The wall bordering the enclosure should be sunk at least 50cm below the ground, be smooth and at least a metre high. An inturned lip along the top of the perimeter wall is advised. Loose soil or, preferably, a pile of loose vegetation cuttings should be made available during the cooler months, beneath which the tortoises will bury themselves while hibernating.

Adults may be, and juveniles should be kept in indoor aquaria or aqua-terraria with the water temperature being maintained at 22-26°C. There must be plenty of room in the aquatic portion of the enclosure for all tortoises within to manoeuvre about freely. A platform or equivalent must be made available, upon which the tortoises may bask under a radiant heat source. Unless the tank receives adequate sunlight through a nearby open window, an artificial source of ultraviolet light will be required, or supplementation of the diet with vitamin D and calcium will be necessary.

FEEDING: Macquarie tortoises will readily feed in the water upon raw meat, insects (and other invertebrates), small mice, small whole fish, etc., as well as plant material. Raw meat should be supplemented with a mixture of calcium and multiple vitamins. Live fish can be introduced to a tortoise pond as a supplemental food source. Uneaten food should not be left to foul the water.

CAPTIVE BEHAVIOUR: Although timid when first collected, this species adjusts well to captivity.

BREEDING: Adult Macquarie tortoises can be readily sexed by the comparative size of the tail. Males have much longer and broader tails than females, despite the fact that overall they don't grow as large. Mating occurs in the spring, with nesting taking place in November or December. As many as 18 eggs may be laid (usually buried near the tortoise's pond), and these should hatch and emerge between two and a half and three months later. Both nesting and emergence of the young seem to be triggered by recent rain, which makes digging much easier. The hatchlings need to be moved to an isolated facility immediately or they may fall prey to the adults.

SELECTED REFERENCES

(Note: if any of the following are unavailable at your public library, most can be borrowed from the libraries of herpetological societies)

Banks, C. (1980) *Keeping Reptiles and Amphibians as Pets*, Nelson, Melbourne, 129pp

Bustard, H. R. (1970) *Australian Lizards*, Collins, Sydney and London. 162pp.

Cann, J. (1978) *Tortoises of Australia*, Angus and Robertson, Sydney, 143pp

Cogger, H. G. (1967) *Australian Reptiles in Colour*, A. H. & A. W. Reed, Sydney, 112pp.

Cogger, H. G. (1980) *Snakes*, Science Field Guide Series: Longman Cheshire, Sydney, 38pp

Cogger, H. G. (1986) *Reptiles and Amphibians of Australia*, Reed Books, Sydney, 688pp, 4rd revised edition

Cooper, J. E. and O. F. Jackson (eds.) (1981) *Diseases of the Reptilia*, Academic Press, London. 2 Vols

Davey, K. (1970) *Australian Lizards*, Lansdowne Press, Melbourne, 111pp

Frye, F. (1973) *Husbandry, medicine and surgery in captive reptiles*, V.M. Publishing Inc., Kansas, 140pp

Goode, J. (1967) *Freshwater Tortoises of Australia and New Guinea*, Lansdowne Press, Melbourne, 154pp

Goode, J. and J. Cann (1974) *Practical Nature Study and Collection*, Sun Books, Melbourne (also published as *Reptiles, Freshwater Animals and You*).

Goin, C. J., O. B. Goin, and G. R. Zug (1978) *Introduction to Herpetology*, W. H. Freeman and Co., San Francisco, 378pp, 3rd edition.

Gow, G.F. (1983) *Snakes of Australia*, Angus and Robertson Sydney, 158pp, revised edition.

Gray, R. L. (ed.) (1985) *Captive Propagation and Husbandry of Reptiles and Amphibians*, Northern California Herpetological Society & Bay Area Amphibian and Reptile Society Joint Publication, U.S.A., 165pp

Gray, R. L. and M.D. Bumgardner (eds.) (1984) *Proceedings of the Northern California Herpetological Society's Conference on Captive Propagtion and Husbandry of Reptiles and Amphibians*, Bulletin of the Chicago Herpetological Society 19(1-2) and the Northern California Herpetological Society Special Publication No. 2

Grigg, G., R. Shine and H. Ehmann (eds.) (1985) *Biology of Australasian Frogs and Reptiles*, Surrey Beatty & Sons, Sydney, 527pp

Halliday, T.R. and K. Adler (eds). (1986) *The Encyclopaedia of Reptiles and Amphibians*, George Allen and Unwin, London. 143pp.

Heatwole, H. (1976) *Reptile Ecology*, University of Queensland Press, St Lucia, Qld, 178pp

Kauffield, K. (1967) *Snakes: The Keeper and the Kept*, Doubleday and Co. Inc., New York, 248pp

Marcus, L.C. (1981) *Veterinary Biology and Medicine of Captive Amphibians and Reptiles*, Lea and Febinger, Philadelphia.

Mattison, C. (1982) *The Care of Reptiles and Amphibians in Captivity*, Blandford Press, Poole, U.K.

McCracken, H.E. (1988). *Husbandry and Diseases of Captive Reptiles*. pp827-878 in Australian Wildlife. Proceedings 104, Postgraduate Committee in Veterinary Science, University of Sydney.

McPhee, D.R. (1979) *The Observer's Book of Snakes and Lizards of Australia*, Methuen, Sydney, 157pp

Mirtschin, P. and R. Davis (1982) *Dangerous Snakes of Australia. An Illustrated Guide to Australia's Most Venomous Snakes*, Rigby, Adelaide, 207pp

Murphy, J. B. and J. T. Collins (eds.) (1980) *Reproductive Biology and Diseases of Captive Reptiles*, Publication of the Society for the Study of Amphibians and Reptiles, U.S.A.

Parker, H. W. and A. G. C. Grandison (1977) *Snakes: a Natural History*, University of Queensland Press, St Lucia, 124pp, 2nd edition

Ross, R. A. (1978) *Python Breeding Manual*, Institute for Herpetological Research, Stanford, California, 51pp

Ross, R.A. (1984) *The Bacterial Diseases of Reptiles*, Institute for Herpetological Research, Stanford, California, 120pp

Schmida, G. (1985) *The Cold-Blooded Australians*, Doubleday, Lane Cove, 208pp

Shine, R. (a wide assortment of ecological papers dealing with the life history of various Australian snakes, bound and held by Australian Herpetological Society library)

Storr, G.M., Smith, L.A. and Johnstone, R.E. (1981) *Lizards of Western Australia I. Skinks*, University of Western Australia Press with Western Australian Museum, Perth 200pp

Storr, G.M., Smith, L.A. and Johnstone, R.E. (1983) *Lizards of Western Australia II. Dragons and Monitors*, University of Western Australia Press with Western Australian Museum, Perth 114pp

Storr, G.M., Smith, L.A. and Johnstone, R.E. (1986) *Snakes of Western Australia*, University of Western Australia Press with Western Australian Museum, Perth 187pp

Sutherland, S. (1981) *Venomous Creatures of Australia*, Oxford University Press, Melbourne, 128pp

Sutherland, S.K. (1983) *Australian Animal Toxins*, Oxford University Press, Melbourne, 527pp

Swanson, S. (1987) *Lizards of Australia*, Angus and Robertson, Sydney, 160 pp revised edition.

Wilson, S. and D. Knowles (1988) *Australian Lizards and Snakes*, A complete photographic and descriptive Index, In Press, Collins, Sydney, est. 700pp